AUSTRALIAN HIGH COUNTRY

RAPTORS

For Sue, with love

AUSTRALIAN HIGH COUNTRY

RAPTORS

JERRY OLSEN

CSIRO PUBLISHING

National Library of Australia Cataloguing-in-Publication entry

Olsen, Jerry, author.

Australian high country raptors / Jerry Olsen.

9780643109162 (paperback)
9780643109179 (ePDF)
9780643109186 (epub)

Includes bibliographical references and index.

Birds of prey – Australia.
Birds of prey – Australia – Identification.
Mountain animals – Australia.

598.910994

Published by
CSIRO Publishing
36 Gardiner Road, Clayton VIC 3168
Private Bag 10, Clayton South VIC 3169
Australia

Telephone: [+613] 9545 8555
Local call: 1300 788 000 (Australia only)
Fax: +61 3 9662 7555
Email: csiropublishing@csiro.au
Website: www.publishing.csiro.au

Front and back cover photos by Julian Robinson

Photographs are by the author or Susan Trost unless noted otherwise

Line drawings by Nicolas Day and Robert Bartos

Set in 11/13.5 Minion and Helvetica Neue
Edited by Anne Findlay, Editing Works Pty Ltd
Cover design by James Kelly
Typeset by Desktop Concepts Pty Ltd, Melbourne
Index by Bruce Gillespie
Printed by Ingram Lightning Source

Jan26_RP_ILS

Contents

Preface and acknowledgements

This book covers raptor species that regularly breed in the high country above 600 m, from Goulburn down to the hills outside Melbourne, as described by the book *Birds in the Australian High Country* edited by Harry Frith (1969). We explore the nature of raptors in general and compare these high country raptors to lower-elevation breeders. Several overseas raptors are described, especially from North America and Europe. It is important to understand these overseas species to determine whether the overseas protocols, models and study designs actually fit Australian raptors.

For the purposes of the book, we refer to *birds of prey* or *raptors*, as animals classified as Accipitriformes, which means 'hawk-shaped birds' (diurnal hawks, kites and eagles), Falconiformes ('falcon-shaped birds') and Strigiformes (nocturnal owls). We begin with a description of *habitats* and vegetation types in the high country, and which raptors are likely to be seen in each habitat type. Next we discuss tips on *finding raptors* and some *basics of raptor identification*. After some protocols on *watching raptors* we discuss *designs for survival* that separate them from other birds, and which allow different *hunting* styles that can help discriminate different birds of prey. Raptors can also be separated by the prey they choose: Australian Hobbies chase birds, Black-shouldered Kites take mice, Wedge-tailed Eagles take kangaroos, so a section on raptor *food* logically follows. Next we cover raptor *breeding and behaviour* which leads to the question *Why are female raptors bigger than males?* followed by a chapter on *raptor conservation*. Appendix 1 gives *species accounts* for breeding species in the high country with basic information about their ecology, distribution and conservation, Appendix 2 gives detailed instructions about handling an *injured or orphaned raptor*.

A great many people contributed to this book and the studies it covers. I would like to thank Susan Trost, Will Osborne, Stephen Debus and Ian Fraser for reviewing earlier drafts. Thanks to numerous people for field assistance and advice including Esteban Fuentes, Rod Kavanagh, Jim Duncan, Tracy Fleming, Jim Hone, Tony Tucker, Nancy Fitzsimmons, John Young, Tom and Gillian Bellas,

Jeanine Lisle, Susan Myers, Greg Hayes, Dalice Trost, Deb Stone, Deana Withey, David Mallinson, Mark Osgood, Colin MacLaren, and Frank and Tanya Barnes. ACT Parks and Conservation gave access to ACT Nature Parks and Reserves and Namadgi National Park, and I am especially grateful to Michael Maconachie, Lois Padgham, Odile Arman, Steve Taylor, Christie Gould, Murray Evans, Brett McNamara, Paul Higginbotham, Kate Boyd, Meg Doepel, Simon Tozer, Darren Rosso, Oliver Orgil, Nick Webb, John McRae, Beth David and David Shorthouse. David Drynan and Lisa Hardy of the Australian Bird and Bat Banding Scheme supplied ABBBS bands and Vic Hurley generously supplied colour bands for some of the Boobook studies. Thanks to David Judge for statistical advice and to Greg Flowers, Joe McAuliffe and Robyn Lawrence for help with the ACT Powerful Owl study. A number of people shared their particular area of expertise including Ed McNabb, Raylene Cooke, Todd Soderquist, Chris Pavey, Rohan Bilney, Matthew Stanton, Paul McDonald, Simon Cherriman, Greg Clancy, Bene Rodriguez, Felipe Siverio, Paul Peake, Tony Rose, Bill Mannan, Liz Urban, Terry Dennis, Jeff Jolly, Paul Willcock, Yvonne Sitko, Nick Mooney, David Bird, Marcos Moleón, Erkki Korpimäki, Jesus Lemus, Joan Real, Luis Palma, Sergio Fabrizio, Vincenzo Penteriani, Michael Wink, Iñigo Zuberogoitia, Lander Astorkia, Inaki Castillo, Claus and Ingrid König, Jan Ove Gjershaug, Torgeir Nygård, Karl Otto Jacobsen, Alv Ottar Folkestad, Per and Lucy Nøkleby, Carl Knoff, Dick Schodde, Ian Mason, Robert Palmer, Joni Downs, Denver Holt, Erik Powell, Kate Davis, David Anderson, Bud Anderson, Jonny Groves, Lloyd Kiff, Ray Cromie, Allen Fish, Jeff Marks, Geoff Holroyd, Gordon Court, Cheryl Dykstra, Victor Escobar, Anir Munoz, Eduardo Navarro, Maria José Moreno, Franscisca Izquierdo, Conseulo Foerster, Miguel Saggese, Colin Trainor, Les Boyd, Walter Boles, Moira Pryde, Leo Joseph, Leah Tsang, Greg Czechura, Ron Johnstone, Wayne Longmore, Philippa Horton, Noel Hyde, Debbie Stewart, Rick Webb, Michael Mulvaney, Les Christidis, Sam Banks and Ken Green. Particular thanks to members of the Institute of Applied Ecology at the University of Canberra who provided logistic help and discussion especially David Choquenot, Arthur Georges, Charles Krebs, Stephen Sarre, Felicity Hatton, Brian Cooke, Bernd Gruber, David Williams, Sue Briggs, the late Richard Norris, Karen Mow, Barbara Harris, Amy Dobos, Jayne Lawrence and Jo Shevchenko. Members of the Canberra Ornithologists Group, including Geoffrey Dabb, Anthony Overs, Michael Lenz, Susan Robertson, Chris Davey, Barbara Allan, Richard Allen, Mark Clayton, Con Boekel, Steve Holliday, Philip Veerman and Roger Curnow provided invaluable help. Thanks also to CSIRO Publishing especially Tracey Millen, John Manger, Melinda Chandler, as well as Anne Findlay for their professional advice and assistance throughout the process of publishing this book.

Particular thanks to Nic Day for his expert drawings used in the main text, Robert Bartos for his detailed drawings used in the main text and Appendix 2, Julian Robinson for sourcing, providing and editing images for the entire book, and Alan Nicol for editing images. Thanks also to Geoff Dabb, Felicity Hatton, George Eade, Charles Davis, Mark Osgood, Michael Maconachie, Kym Bradley, Roger Williams, Jesús Bautista, José M. Gil-Sánchez, Derek Willcox, David Pastern and Mark Sweany who supplied images.

1

The high country

The nest may have dated back 40 years, to the days of Leopold and Wolfe. We were standing on a platform of sticks up in a Scribbly Gum. Tools clinked as we laid them out, and sun glinted off the black beak and talons of the eagle. An eagle nest is big enough to use as a workbench, as long as you work carefully. Esteban and I preferred to handle and band the young eagle up in the nest (Figure 1.1) rather than down on the ground where it was hot and dead still. Breeze fluttered through the downy white feathers on the eagle's head (Figure 1.2), and cooled the sweat on our backs from the walk up the hill and climbing the rope. Above us two black adults soared in tight circles against blue sky (Figure 1.3), watching us and then rocking in the air from side to side as vultures do. A pair of ravens came in hard from the south; with quavering calls, they dived at the eagles. The female Wedgie closed her wings and sailed down, pinions back, lowering her undercarriage, feathered legs and black talons thrust forward, as she landed on a dead tree.

Around the rim of this nest, in the gently rocking tree, lay castings and pieces of animal – kangaroo, rabbit, a fox leg, cockatoo feathers. We labelled and bagged these for Tony Rose, the retired dairy farmer on the New South Wales coast, who worked magic with his microscope and private animal collection to identify dozens of bird, mammal and invertebrate species from castings and prey remains. Tony has identified prey species that weren't even supposed to be in the ACT, like bandicoots and jaegers.

Figure 1.1 Esteban Fuentes climbing to a Wedge-tailed Eagle nest.

This nest could have been one studied by Starker Leopold and Tony Wolfe almost 40 years earlier when serious raptor study began in Australia. It began in the high country.

If you study birds of prey in Australia, 1969 was an important year. Harry Frith, Chief of the CSIRO Division of Wildlife and Rangelands Research, teamed up with an artist, Betty Temple Watts, to publish *Birds in the Australian High Country.* John Calaby, a veteran field biologist, penned the raptor accounts for *High Country.* In these species accounts we get one of the first comprehensive and scientific treatments of about half of Australia's birds of prey. The book remains a classic and is still consulted four decades later.

The other seminal event was an eagle paper submitted to the journal *CSIRO Wildlife Research* by A. Starker Leopold and Tony Wolfe, the first major scientific study of any Australian raptor. This study gave a kick-start to serious raptor research in Australia. Papers on Australian raptors were published in journals such as *Emu*, and David Fleay had studied and written about Powerful Owls in the high country, but it was Leopold and Wolfe who conducted the first statistical analysis drawn from large sample sizes near Canberra. They measured the food habits of Wedge-tailed Eagles on the elevated tablelands outside the city. Starker Leopold

Figure 1.2 Nestling Wedge-tailed Eagle from the Leopold and Wolfe study area, 40 years later.

Figure 1.3 High country Wedge-tailed Eagle (Julian Robinson).

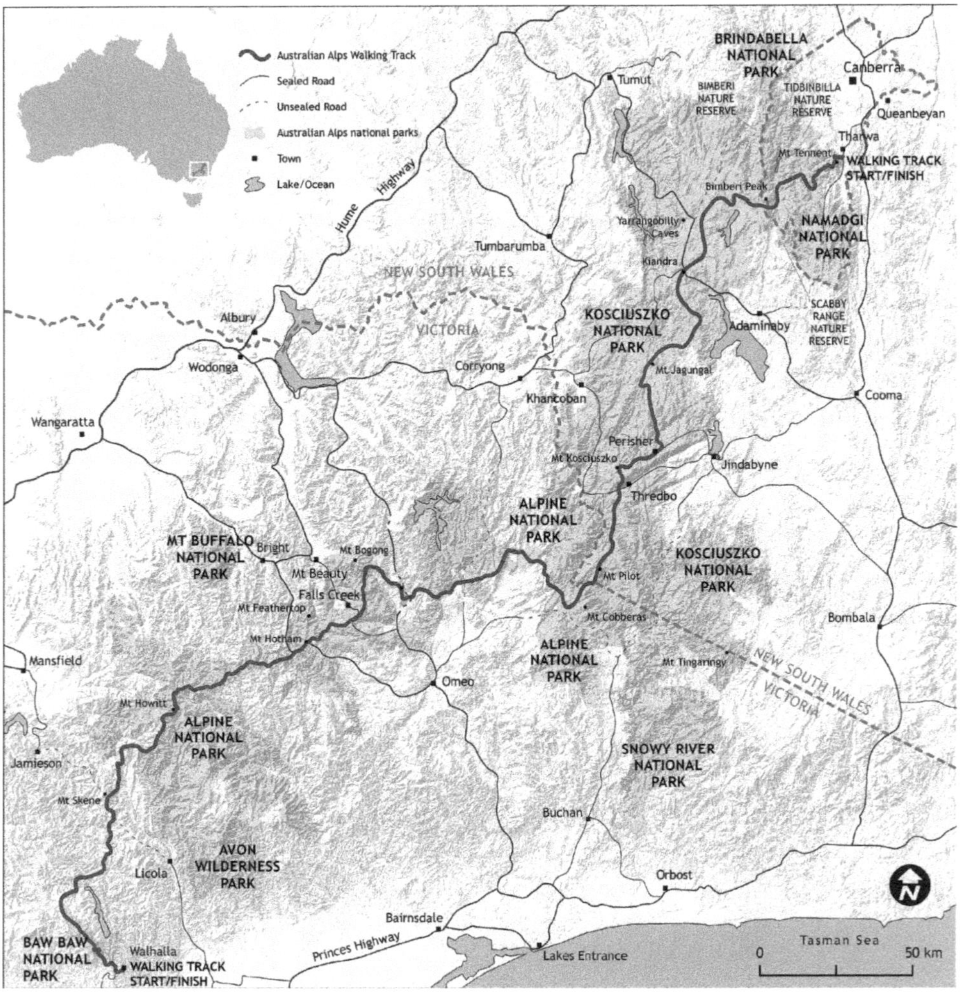

Figure 1.4 The Australian high country (ACT Parks and Conservation).

was based at the Museum of Vertebrate Zoology at the University of California in Berkeley. He had a good pedigree – he was the son of the renowned ecologist Aldo Leopold who wrote *A Sand County Almanac*. Wolfe was a technical officer with CSIRO's Wildlife Department.

These high country studies, *Birds in the Australian High Country*, and Leopold and Wolfe's 'Food habits of nesting Wedge-tailed Eagles, *Aquila audax*, in south-eastern Australia' (1970) influenced many raptor studies that followed.

The high country described by Frith (1969) extends from near Goulburn on an elevated plain, ranging from just below 600 m (metres above sea level (masl)), up the slopes and spurs to Mt Kosciuszko at 2228 m, and south through the Snowy Mountains to the Victorian Alps (Figure 1.4). These remnants of an old mountain

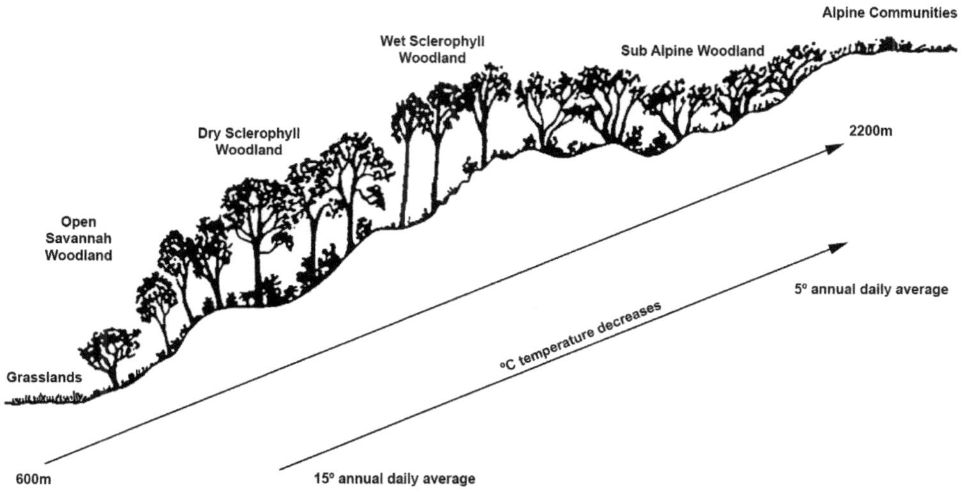

Figure 1.5 Altitudinal stratification in the high country (after Mason undated)

system were eroded to a low level then uplifted in parts, but are nowhere as high as the mountains found in North America or Europe. In places the region is intruded by granite masses and great sheets of lava that poured down from ancient volcanoes. Streams have cut gorges and broad valleys through this country. It is dominated by *Eucalyptus* forest and woodland, a key to understanding Australian raptor biology. The exceptions are swamps, the city of Canberra, towns, pine plantations, grasslands and alpine heaths. The main raptor habitats show a distinct altitudinal stratification in type from the low valleys up to the tops of the mountains – grasslands, lowland eucalypt woodland, mixed eucalypt forest, Snow Gum subalpine woodlands, alpine heathland and feldmark (see Figure 1.5).

The relatively flat Monaro Plain runs south from Canberra to Cooma, uplifted, dissected and eroded, leaving a high plateau. The ranges and mountains in the high country aren't the craggy pinnacles found in Europe or western North America, but feature large rounded masses of granite and sedimentary material, the lower ones used by nesting Peregrines (in this book 'Peregrine' in an Australian context refers to *Falco peregrinus macropus*) and kestrels, the higher ones avoided by all breeding raptors. Mt Kosciuszko is not really a peak but the highest point of a gently rising system of ridges.

Very little of this country lies below 600 m and much of it sits above 900 m; the climate differs in the lower valleys compared with high plains and mountain tops, and by Australian standards the weather is generally harsh throughout.

Extreme temperatures in Canberra from 1939 to 2010 ranged from 42.7°C to –10.0°C. In the high mountains near Charlotte Pass (1760 m) extremes from 1930 to 2012 were 26.7°C and –22°C. Frosts are common across the high country. In the alpine areas the mean minimum temperature remains below 0°C for 6–8 months each year. Above 1230 m snow falls every year; in parts of the alpine region it can lie for most months of the year. The average rainfall varies greatly, from 450 mm in Cooma to more than 2 m in some high mountain areas. The El Niño Cycle is significant, and in some years both summer and winter rainfall fail causing severe droughts, tree mortality and drying up of water sources. In general the cycle of plant growth is regular and prevented by cold temperatures in winter, or by drought and high evaporation in summer. In spring and early summer there is abundant plant growth and most raptors breed then, though Wedge-tailed Eagles and Powerful Owls initiate breeding before mid-winter, and Black-shouldered Kites on the tablelands often breed twice, once in autumn, and again in spring if their rodent prey is abundant.

Figure 1.6 Bullen Range Wedge-tailed Eagle nest after the 2003 fire.

Aboriginal burning probably changed high country vegetation (e.g. Gammage 2012), and European settlement changed it further. This human intervention – including stock grazing – altered the composition of grasslands, created lakes and dams, destroyed swamps and woodlands, cleared or thinned native forest, often through ring-barking, and replaced some with exotic pines. Domestic grazing and feral animals altered the understorey of some forests. Changes were greatest on the plains and ridge tops and least in the mountains where vast areas of timber remain in relatively virgin condition, though fires have changed much of this forest in the last decade.

At the centre of the high country the Australian Capital Territory (ACT) covers 2360 km^2 of elevated plains and mountains with a diversity of raptor habitats. However, these are not distinct from the surrounding New South Wales and nearby Victorian habitats. Raptors that breed in the high country of New South Wales and Victoria also breed in the ACT, although Sooty Owls, Masked Owls, Barn Owls and Grey Goshawks have not been found breeding in the ACT.

The soils in the highlands are generally poor and this influences vegetation and fauna, including the numbers of breeding raptors there. Higher altitudes are exposed to more wind, cold, rainfall and snow. North- and west-facing slopes receive more direct sunlight so they are drier and warmer than south- or east-facing slopes, and this can affect hunting and nest choices made by breeding raptors. On 18 January 2003 a massive crown fire, started by lightning strikes in the Brindabella Ranges, cut through ACT and New South Wales forests and woodlands (Figure 1.6), including more than 90% of Namadgi National Park in the ACT. Highland Victorian forests are more degraded by logging and they were further affected by fires in 2009. These burnt forests and woodlands are now recovering and this affects the breeding of certain raptors.

2

Comparing Australian raptors to their Northern Hemisphere counterparts

Many of the protocols used to study Australian raptors, and the principles used to analyse Australian raptor studies, are drawn from North America and Europe, though our raptors and their prey differ in significant ways from Northern Hemisphere raptors and their prey. Sometimes we fail to understand the adaptive behaviours of Australian raptors because we're myopic, confined and limited by overseas studies and trying to fit our species to patterns found overseas. However, contrasting our raptors with their closest relatives overseas can be helpful. Throughout the text I'll point to certain similarities and differences between Northern Hemisphere raptors and their congeners in Australia at similar latitudes, around 36° north and south of the equator. We can start with two interesting contrasts that involve habitat: **1) Cliff nesting** and **2) Upper altitudinal breeding limit.**

What do Australian raptors and corvids have against cliffs?

Travelling through the Sierra Nevada mountains of southern Spain with the raptor expert Marcos Moleón, over winding dirt lanes through lovely grassland country scarred with bare rock, we stopped and walked up a gully to a short cliff. Through binoculars we examined a stick nest ~10 m up on a rock face. This nest could have

Figure 2.1 Wedge-tailed Eagle nest in a White-bellied Sea-Eagle cliff nest near Port Lincoln, South Australia (Jeff Jolly).

been built by any number of species of eagles, hawks or corvids. Northern Hemisphere *Aquila* eagles such as the Golden Eagle and Bonelli's Eagle, and *Hieraaetus* eagles such as the Booted Eagle, as well as *Corvus* species such as the Common Raven, build nests on cliff ledges and in potholes on cliffs. Falcons cannot build a nest so Peregrines, Gyrfalcons, Sakers, Luggers, Lanners and Prairie Falcons use the nests of Rough-legged Hawks, Common Ravens, Golden Eagles and other raptors that do build nests on cliffs (Cramp and Simmons 1980; Johnsgard 1990). But Australian *Aquila* and *Hieraaetus* eagles, and corvids such as the Australian Raven, don't build nests on cliffs. They don't seem to know how, though on the Australian coast, White-bellied Sea-Eagles and Ospreys do use rock pinnacles and rock faces and Forest Ravens and Little Ravens occasionally build on island cliffs. Wedge-tailed Eagles will use cliff nests on the coast built by sea-eagles (Figure 2.1).

Inland falcons do not then have the luxury of a stick nest on a cliff. They can nest in a tree nest or tree hollow, on bare rock, or on a building but that's it. Close relatives of Northern Hemisphere falcons, the Australian Brown Falcon, Grey Falcon and Black Falcon never nest on cliffs. Unlike in Spain, Utah or California, inland Australian Peregrines and Nankeen Kestrels have cliffs to themselves. In southern Spain Peregrines share cliffs with Golden Eagles, Booted Eagles, Bonelli's Eagles, ravens or Common Buzzards and, unlike Australian Peregrines, Spanish Peregrines often have a comfortable stick nest built by a hawk or eagle to cushion

Figure 2.2 Marcos Moleón showing us a Bonelli's Eagle cliff nest near Granada in Spain.

their eggs. Why is Australia so different? Why don't eagles, ravens and other raptors build nests on cliffs (see Figures 2.2 to 2.5) like their closely related Northern Hemisphere counterparts do? Have they lost this ability, or didn't they have it in the first place?

Perhaps geology is the answer; Australian cliffs are older, more worn down. Are they less suitable for Australian raptors because they are smaller or different? This doesn't stand up to scrutiny. In the Blue Mountains, throughout Namadgi and Kosciuszko National Parks and elsewhere in Australia there are plenty of cliffs and cliff ledges big enough, but eagles and ravens don't use them. Furthermore, I've climbed into Golden Eagle cliff nests in Idaho, USA, and Alberta and the Yukon in Canada, without a rope, all on low, accessible cliffs that even Australian Peregrines or kestrels would be reluctant to use. If you climb into enough Australian Peregrine nests you see plenty of ledges big enough to hold an eagle nest and many ledges big enough to hold a large motorcar. Geology seems an obvious reason, but it doesn't add up. Maybe we should look at trees.

In 2009 Allen Fish, the Director of the Golden Gate Raptor Observatory, drove us to the Presidio overlooking the Golden Gate Bridge in San Francisco looking for Red-shouldered Hawks, the best place, he said, to find them. We walked through

Figure 2.3 Nestling Bonelli's Eagles on nest built on a cliff.

Figure 2.4 Peregrines in Spain can use stick nests built on cliffs by other raptors (Iñigo Zuberogoitia).

Figure 2.5 Peregrines in inland Australia have no stick nests on cliffs to use, so they lay their eggs on bare soil or vegetation.

Figure 2.6 California Golden Eagle nest in eucalypt (George Eade).

Figure 2.7 Eagles in southern Spain, for example, at Estación Biológica de Doñana, near Seville, Spain, often use mature eucalypts for nest trees.

linked woodlands of giant introduced Australian eucalypts, shredded bark crunching underfoot, and he showed us nesting Red-shouldered Hawks, Cooper's Hawks, Red-tailed Hawks and Great Horned Owls. The following day we crossed the Golden Gate Bridge to Sausalito and saw more Red-shouldered Hawks and Red-tailed Hawks, so common that Allen used the local code 'tail' or 'shoulder' as shorthand when he spotted one. Swainson's Hawk, and White-tailed Kites (the equivalent of our Black-shouldered Kites), American Kestrels, more Great Horned Owls, all nested in Australian eucalypts. On a ranch above Sausalito, a Golden Eagle (Figure 2.6) nested in a Southern Blue Gum *Eucalyptus globulus* looking remarkably like a Wedge-tailed Eagle nest with downy young. All of these American raptors were wedded to Australian eucalypts.

A curious thing when you visit Spain, Portugal, California, Arizona, and other places is that local raptor biologists, keen to show you nesting eagles and hawks, take you to groves of eucalypts (Figure 2.7). In Portugal Luis Palma took us to Bonelli's Eagle nests and almost all were in introduced Australian eucalypts; according to Luis, they seem to prefer these over smaller native oaks or pines. Similarly, Bill Mannan in Tucson, Arizona, took us to several Cooper's Hawk nests mostly in eucalypt trees, these hawks' preference (Figure 2.8), and a Great Horned Owl's nest (Figure 2.9). Even the rare Imperial Eagle in Spain, among other species (Figure 2.10), nested in eucalypts around Seville. Raptors nest in eucalypts in

Figure 2.8 Cooper's Hawk nest in eucalypt in Arizona (Liz Urban).

northern Chile but rarely around Santiago (Victor Escobar pers. comm.) so this pattern is not perfect; however, we probably should consider eucalypts as a possible part-answer to the puzzle of why so many Australian raptors avoid cliffs as nest sites. Maybe in our eucalypt-dominated Australian landscape we have superior nest trees, so cliffs are unnecessary. This could also explain why Barn Owls in Australia (see below) almost never nest in buildings, preferring trees instead, while their counterparts in the Northern Hemisphere and Indonesia commonly do. A protocol for studying Prairie Falcons in the United States is 'Don't kill the Barn Owls'. Any US falcon researcher would understand what this means, but it's fairly obscure to Australians. In the US Barn Owls are cliff-nesters, and when flushed from their nest ledge, or pothole nest in a cliff, defending Prairie Falcons swoop and kill them. It's a common problem in the US, but not a problem for Peregrine researchers in Australia. You can stand on the ground in the US and look up at a brooding or incubating Barn Owl in a limestone cliff. Bunn *et al.* (1982) said

Figure 2.9 Great Horned Owl guarding its nest in a eucalypt in Tucson, Arizona.

European Barn Owls started out as cliff-nesters, then moved to buildings, with some nesting in hollow trees.

If you walk into limestone canyons on the Indonesian islands of Sumba or Roti, you find the same thing – Barn Owls nest in potholes in cliffs, rather like Peregrines do in Australia, and you commonly find them nesting in churches and other buildings throughout these islands. This could explain why Barn Owls in Australia almost never nest in buildings, preferring trees instead, while their counterparts in the Northern Hemisphere and Indonesia often nest in buildings – these foreign Barn Owls originally evolved mainly as cliff-nesters (Bunn *et al.* 1982; Johnsgard 2002). The shift to buildings was a closer match for the more northern species, whereas Barn Owls in Australia evolved to nest mainly in eucalypts so the shift to buildings was ecologically distant.

Another reason why Australian raptors avoid cliffs could be that Australian raptors and corvids have been isolated for so long from their Northern Hemisphere congeners, by Wallace's Line and other geographic barriers, that Australian raptors have lost the ability to build nests on cliffs, or during this separation their

Figure 2.10 Bonelli's Eagle nest in eucalypt in southern Spain (Jesús Bautista).

Northern Hemisphere counterparts learned it. This topic is worthy of study because *Falco* species such as Brown Falcons, Australian Hobbies, Grey Falcons and Black Falcons compete for these limited tree nests. My guess is that Brown Falcons, a particularly scrappy species, would win many of these contests, to the detriment of the much rarer Grey Falcons and Black Falcons, even though Black Falcons are heavier, and on paper, should be stronger.

Upper elevation breeding limits – are Australian raptors sissies?

The village of Patagonia in southern Arizona is a paradise for birdwatchers because it is like a shard of Mexican habitat poking up into the American underbelly. Many Mexican species occur in these mountains and rolling hills that you can't see

Figure 2.11 Peregrine near its nest at ~1750 m near Patagonia, Arizona.

elsewhere in the US, like the beautiful Gray Hawk. Coming down out of the hills through oak woodlands, the road levels out into town. The town doesn't look elevated, but the main street is as high as any Australian Peregrines ever nest – 1200 m. Erik Powell, a raptor expert from the area, drove us out of town into rocky hills along the Mexican border to a Peregrine nest on a cliff at ~1750 m elevation, higher than any known Australian Peregrine nest and maybe higher than any Australian raptor nest ever found (Figure 2.11). Santa Fe, capital of New Mexico at 2134 m is much higher, almost as high as the top of Mt Kosciuszko, but such elevations are not too high for breeding Peregrines, eagles or hawks in North America or Europe. At around 36° north in the United States, Peregrines in Colorado breed at 3200 m, in California at 3600 m. The 1200 m Australian cap is about the same upper limit you find for Peregrines breeding in Scotland, 56° north and famous for its inclement weather. Are Australian Peregrines sissies?

Northern Hemisphere Spotted Owls breed at 2700 m, but Powerful Owls haven't been found breeding above 1200 m in Australia. Booted Eagles in Europe breed at 1500 to 2000 m, Bonelli's Eagles at 1500 m, Golden Eagles in chilly Switzerland breed at 2400 m, and around Granada, in the Sierra Nevada (Andalucía) of Spain at 2600 m, but their close relatives the Wedge-tailed Eagle and Little Eagle never breed above 1200 m. North American Prairie Falcons, which resemble their close

Figure 2.12 Banding nestlings in the Brindabella Ranges near the 1200 m upper limit for breeding Peregrines in Australia.

Australian relative the Brown Falcon and live on many mammals, breed at 3700 m in Colorado, but Brown Falcons don't breed above 1200 m.

The upper altitudinal limit for raptors breeding in the Australian high country is less than half the altitude of those breeding at similar latitudes in places like Spain, California or Arizona. Sometimes you see Australian Peregrines and Wedge-tailed Eagles flying in the altitudinal band between 1200 m and the top of Mt Kosciuszko at 2228 m, but not breeding at that altitude. For Peregrines the 1200 m breeding limit is remarkably consistent across the Australian high country (Figure 2.12); in Tasmania it is closer to 900 m (Table 2.1). There is something harsh about these high elevation habitats that prevents raptors from breeding. In the ACT, Peregrines' home ranges at around 1200 m are larger and reproduction is poorer than for pairs breeding down at 600 m (Olsen *et al.* 2004), even though there are many rocky outcrops and cliffs above 600 m.

What is it about these higher elevations in Australia that makes them unsuitable for breeding? A knee-jerk response would be food, or lack of it. The

Table 2.1 Upper elevations and type of nest used (tree or cliff) for some breeding species in south-east Australian high country and their counterparts elsewhere in the world

Species	Region	Upper elevation (metres)	Cliff or tree nesting
Peregrine	ACT	1200	cliff + tree
Peregrine	New South Wales	1200	cliff + tree
Peregrine	Victoria	1190	cliff + tree
Peregrine	Tasmania	900	cliff
Peregrine	Colorado	3200	cliff
Peregrine	Scotland	1200	cliff
Brown Falcon	ACT	1000	tree
Prairie Falcon	Colorado	3700	cliff
Australian Hobby	ACT	1000	tree
Brown Goshawk	ACT	1000	tree
Northern Goshawk	New Mexico	2530	tree
Cooper's Hawk	New Mexico	2380	tree
Sharp-shinned Hawk	Colorado	3200+	tree
Wedge-tailed Eagle	ACT	1000	tree
Golden Eagle	Spain	2600	cliff
Golden Eagle	New Mexico	2900	tree + cliff
Little Eagle	ACT	800	tree
Booted Eagle	Spain	2000	tree + cliff
Bonelli's Eagle	Spain	1500	tree + cliff
White-bellied Sea-Eagle	New South Wales	1200	tree
Bald Eagle	New Mexico	2650	
Black-shouldered Kite	New South Wales	1000	tree
White-tailed Kite	Arizona	1200	tree
Osprey	New South Wales	100	tree
Osprey	New Mexico	2130	tree
Powerful Owl	ACT	1120	tree
Great Horned Owl	New Mexico	2530+	tree + cliff

alpine country on Mt Kosciuszko looks remarkably like alpine country in Colorado or Europe but there are differences. The latter has gallinaceous birds such as the ptarmigan *Lagopus*, marmot *Marmota*, ground squirrel *Spermophilus* and hares *Lepus* species, that stay year-round above 1400 m in southern Europe and Colorado providing food for raptors such as Prairie Falcons and Golden Eagles there. We have no counterparts in the Australian high country. There is some evidence for this lack of prey theory. Australian Peregrines nesting at the 1200 m upper limit often kill and eat ravens, but high country Peregrines down around 600 m do not.

Table 2.2 Geometric Mean Prey Weight compared to the average weight of Peregrines and prey/predator index of Australian high country Peregrines compared with other Peregrines

Peregrine subspecies	Location	GMPW (g)	Predator weight	Prey/predator ratio
Falco peregrinus peregrinus	Europe	181.4	889	0.20
F. p. anatum	North America	124	781	0.16
F. p. cassini	South America	80.3	800	0.10
F. p. macropus	Australian highlands	202.1	745	0.27
F. p. macropus	ACT, Australia	132.9	745	0.18
F. p. anatum	North America	124	781	0.16
F. p. cassini	South America	80.3	800	0.10

Geometric Mean Prey Weight (GMPW is a calculation of average weight of prey species taken by a raptor) and *predator/prey ratio* (GMPW compared to the average bodyweight of a raptor) for Peregrines at higher Australian altitudes compared to elsewhere in Table 2.2 shows that Peregrines in the Australian high country take proportionally heavier prey than do Peregrines elsewhere, supporting the notion that prey could be scarcer at higher altitudes in Australia.

But this may be simplistic, and answering the question before testing it would be poor science. The explanation could be a combination of factors including food scarcity and plant growth. In Australia the tree line, (the habitat edge where trees stop growing) at 36° south is at ~1800 m. In Chile at 37° south it is at 1600 m, in Tasmania at around 42° it is at 1200 m, in New Zealand at 41° it is at ~1500 m, so there is some consistency for the tree line in the Southern Hemisphere; the further from the equator, the lower the tree line. This is also true for the Northern Hemisphere, with an important difference – the tree line is much higher. The Grand Canyon rim, at the same latitude and same altitude as Mt Kosciuszko (2230 m), is covered in sizable trees and has nesting goshawks and other raptors (Figure 2.13). Arizona and Utah accipiters at around 36° north breed in stands of forest at between 1700 and 3000 m elevation. Coniferous forest begins at lower elevations in Arizona, for Ponderosa Pine at around 1825 to 2130 m depending on slope exposure; the upper limit for Ponderosa Pine is ~2750 m. In the Japanese Alps at 36° north the tree line is at 2900 m (Table 2.3), even though it is said that tree lines are lower on islands than on mainlands. In the Sierra Nevada in California at 37° it is at ~3400 m, in the Sierra Nevada in Spain at 38°, because of low precipitation it is at ~2400 m, in the Himalayas at 28° it is at ~4200 m, in the Wasatch Mountains of Utah at 40° it is at ~2900 m, and in the Rocky Mountain National Park in Colorado at 40° it is at ~3400 m.

Paradoxically, one reason for lower tree lines in the Southern Hemisphere seems to be low ground temperatures during the growing season. According to

Figure 2.13 At 2200 m in Arizona, the same elevation as the top of Mt Kosciusko, you are below the tree line.

Table 2.3 Some tree lines in the Northern and Southern Hemispheres (Sources: Encyclopaedia Britannica World Atlas 2006; Körner and Paulsen 2004)

Location	Latitude	Altitude (metres)
Charlotte's Pass, Australian Alps	36° S	2000
Japanese Alps	36° N	2900
Termas de Chillan, Concepción, Chile	36° N	1600
Sierra Nevada, Spain	37° N	2400
Yosemite, California	38° N	3320
Colorado	38–40° N	3610
Kaimanawa Mountains, Turangi, New Zealand	39° S	1350
High Atlas Mountains, Morocco	35° N	2450

Körner and Paulsen (2004), when you disregard taxon, land-use or fire-driven tree limits, high altitude climatic tree lines around the world are associated with a seasonal mean ground temperature of 6.7°C (± 0.8 s.d.); a surprisingly narrow range for tree lines around the world (Ken Green, pers. comm.). This is linked to the huge expanses of water in the Southern Hemisphere compared with the Northern Hemisphere, as it takes a lot more of the sun's energy to heat water than to heat land. Tree taxa may also play a role; introduced conifers can grow above the tree line in New Zealand, and maybe in Australia, so our dominant eucalypts again could be part of the answer.

Another factor limiting the breeding elevation of Australian raptors could be that soils are poor in the high country and few suitable nest trees grow where Snow Gums dominate. Snow Gums are gnarly and don't produce good hollows for owls or kestrels, and probably there aren't enough invertebrate, small mammal, reptile or bird prey at these elevations. Again, contrasting our raptors with the Northern Hemisphere, Thomas Balgooyen (1976), supervised by Starker Leopold, set his classic American Kestrel study with pairs nesting at between 1700 and 1973 m in California, at around 40° north, a bit like setting a raptor breeding study on Mt Kosciusko. These small falcons took many invertebrates, reptiles, small mammals and birds, prey that may or may not be available to Nankeen Kestrels at higher elevations.

Another difference is that North America has woodpeckers, primary excavators that dig holes in trees in which kestrels nest, even in small trees, smaller than our Snow Gums. Australia has no primary excavators, such as woodpeckers, so above 1500 m kestrels are limited to rock crevices or rare naturally forming hollows in Snow Gums. A pair of Nankeen Kestrels nests in a rock crevice on Mt Coree on the ACT–NSW border at ~1400 m, higher than any Peregrine nest yet found, but their diet has not been studied. Kestrels and Southern Boobooks may be the only Australian raptors that breed above 1200 m.

However, the link between raptor breeding and tree lines isn't consistent. Victor Escobar showed us Peregrines nesting well above the tree line in the Chilean Andes. At least two areas of research are worth pursuing in relation to habitats and elevation: 1) Can we find raptors nesting at higher than 1200 to 1400 m in south-eastern Australia? and 2) Can we determine if a lack of abundant prey, nest habitat, primary excavators and/or frequent hostile weather at these altitudes prevent or destroy nesting attempts, and stop tree growth compared with similar elevations in North America, South America, Africa and Eurasia?

3

Raptor habitats in the high country

Raptors use all habitats in the high country from the elevated tableland plains at 600 m to the slopes and spurs of Mt Kosciuszko at 2228 m, but different raptor species are more common in, and use, some habitats more than others.

City parks, urban gardens – The main urban centre in the high country, Canberra, was designed as an open-space city; extensive planting of trees and shrubs has provided rich, distinctive habitat for birds. Flowering eucalypts, acacias, banksias, grevilleas and numerous exotic berry-bearing shrubs and trees and other plants have diversified the available food and cover for birds such as honeyeaters and currawongs. Most cities around the world are dominated by introduced birds, but Canberra is dominated by native parrots, currawongs and magpies alongside introduced mynas, sparrows and starlings, all major prey species of Peregrines, Brown Goshawks, Australian Hobbies, Nankeen Kestrels and Collared Sparrowhawks which nest in the city (Figure 3.1). Major bird migration routes follow the plantings of the streets and lake foreshores.

Collared Sparrowhawks often nest in introduced pines such as *Pinus radiata* in parks or in people's backyards, and Hobbies (in this book 'Hobby' refers to the Australian Hobby, *Falco longipennis*) choose raven nests in tall eucalypts or sometimes in pines in city parks or backyards. Southern Boobooks select eucalypt hollows in urban parks or along bike paths if these parks are little disturbed, but mostly nest in nature parks and hunt in urban settings after dark. Kestrels nest in eucalypt hollows and on buildings. This tallies with overseas studies where Eurasian Sparrowhawks and Northern Hobbies nest in city parks, often in pines,

Figure 3.1 Collared Sparrowhawk just-fledged from a pine in a city backyard.

and Eurasian Kestrels often nest on high buildings. In North America, Merlins (quick flying, stocky little falcons), American Kestrels (roughly two-thirds the size of ours) and small owls also nest in cities. Even the powerful Eurasian Goshawk nests in city parks in Germany, and Great Horned Owls nest in city parks and front yards in North America. By contrast, large owls in the south-eastern Australian highlands tend not to nest in cities, so you are unlikely to find a Powerful Owl or Masked Owl nesting in an urban front yard. Australian raptors are much shyer than their North American counterparts. The Cooper's Hawk, a close relative, similar-sized and an ecological equivalent of our Brown Goshawk, often nests in cities, even in people's front yards. However, the Brown Goshawk is much more wary, though some are becoming tamer and now nest closer to houses in nature parks. Consequently, findings from surveys of the much-studied Cooper's Hawk need to be treated with caution when applied to the Brown Goshawk.

Wetlands, shores, rivers, lakes and dams – The region has few lowland swamps and many are small and heavily overgrown with sedges and shrubs, shallow and mostly semi-permanent. Littoral habitats in the highlands fringe many dams, rivers and swamplands. Unlike many birds, raptors seldom drink; they get most of their moisture from their carnivorous diet. When they do drink, they do so by scooping water into the lower mandible and swallowing it; unlike pigeons, they cannot suck water. However, they bathe, and the water's edge provides a good place for bathing and a place to ambush prey. It also provides tall trees for nesting, so raptors often breed around water edges.

The south-eastern high country is the source of three major rivers, the Murray, Murrumbidgee and Snowy, and many tributaries. In the upper reaches of these rivers are steep and rocky rivulets and sources of large trees for nesting owls and accipiters. Swift-flowing mountain streams cover the region; stream bottoms are stony or sandy and the banks are often dissected with occasional sand beaches often in eucalypt forest or casuarinas with a dense undergrowth of shrubs, blackberries and ferns. These provide habitat for nectar-feeding birds such as finches, silvereyes and other small passerines, the prey of accipiters and Hobbies, and cover for rabbits taken by species such as eagles, kites and Brown Falcons.

Below the mountains, where rivers run onto flatter ground, are large casuarinas and eucalypts, alive and dead, where Wedge-tailed Eagles often place their massive nests. Brown Goshawks commonly use live riverside casuarinas and eucalypts. Certain species, such as Whistling Kites, White-bellied Sea-Eagles and Swamp Harriers, need water for foraging so are only found nesting close to water, which enables them to efficiently cart prey from the water's edge to the nest. Peregrines and Hobbies prefer to hunt over open rivers and dams where their prey cannot dive into cover, and Peregrines breed on cliffs cut into canyons along rivers and creeks. Swamp Harriers prefer swamps and long, wet grasslands for breeding, for example, near Collector or Hoskinstown, and in Namadgi and Kosciuszko National Parks. These harriers migrate to warmer, low-elevation places for the winter.

There are some small glacial lakes in the high mountains, but the most important lakes are at lower elevations: Lake George, the largest natural lake in New South Wales, and nearby Lake Bathurst, ebb and fill but in recent decades have been dry for extended periods. Water storage for cities, agriculture and hydroelectricity led to the construction of large dams including Lake Eucumbene, Lake Burley Griffin, Googong Dam, Bendora Dam and Corin Dam, among others. These provide habitat for waterbirds, the prey of Peregrines (Figure 3.2), goshawks and Little Eagles; White-bellied Sea-Eagles and Wedge-tailed Eagles prey on the larger waterbirds. In North America Bald Eagles and Ospreys commonly breed far inland on such bodies of water but, so far, ornithologists have not found Ospreys breeding inland in south-eastern Australia and none breed in the high country. Bald Eagles in North America breed at much higher elevations than White-bellied

Figure 3.2 Peregrine nest on the Murrumbidgee River, NSW.

Sea-Eagles do in the highlands (Table 2.1); Sea-Eagles wander to high altitudes but don't breed there. Perhaps the water isn't clear enough for hunting Ospreys, and suitable fish, turtles and waterbirds aren't plentiful enough for Sea-Eagles.

Grasslands – These generally occur below 900 m and were originally dominated by *Themeda, Rytidosperma* and *Poa* species. Native grasslands are now rare in the high country plains, and disappearing. Many original grasslands are now replaced by cereal crops, concreted over in the case of Canberra, planted with trees or 'improved' with introduced grasses such as Phalaris. Some larger surviving patches of native grassland remain in places such as the Monaro Plain and in the Jerrabomberra Valley in the ACT. The change from tall to short grasses could help raptor species such as kestrels. Only the Swamp Harrier nests on the ground in grassland (Grass Owls have not been found breeding in the high country), but many raptors hunt over grassland, especially eagles, Black-shouldered Kites, kestrels, Barn Owls, Brown Falcons and Peregrines (Figure 3.3).

Open savannah woodland – Below 700 m the main forest structure is open grassy eucalypt woodland. Trees rarely exceed 20 or 30 m in height and are widely spaced, giving a parkland appearance. Since European settlement, these trees have been thinned out and the ground cover is composed of degraded native grasslands or pasture of exotic grasses. These woodlands are threatened throughout south-

Figure 3.3 Peregrine nest cliff in NSW grassland.

eastern Australia, where remnant Yellow Box *Eucalyptus melliodora*, Blakely's Red Gum *E. blakelyi* and Red Box *E. polyanthemos* woodland with an understorey of shrubs, herbs and grasses have been thinned for grazing. Many raptors prefer Yellow Box for nesting because these are often the tallest trees. Little Eagles, Brown Goshawks, Brown Falcons and Wedge-tailed Eagles breed and hunt in such open woodlands.

Dry sclerophyll forest – At altitudes of 760 to 900 m the most extensive forest is dry sclerophyll of Red Stringybark *E. macrorhyncha*, Broad-leafed Peppermint *E. dives*, Scribbly Gum *E. rossii* and Brittle Gum *E. mannifera*. These trees rarely exceed 20 m high and can form a closed canopy. Some tall and smaller shrubs, of 1 m or less, often form a more or less continuous layer underneath. This forest is often damaged by fire but on the whole it is closer to its original condition than is the grassy woodland. In the south-eastern highlands there are two main types of dry sclerophyll forest: 1) low altitude Red Stringybark (Figure 3.4), Scribbly Gum (Figure 3.5), Brittle Gum: Brown Goshawks, Collared Sparrowhawks and Little Eagles breed and hunt in these low altitude forests, Brown Goshawks and Little Eagles often nest there in Red Stringybarks on east-facing hillsides. Brown Falcons and Hobbies use the abandoned nests of corvids, and Peregrines take over used corvid and eagle nests; and 2) at higher altitudes Candlebark *E. rubida* and Broad-leafed Peppermint prevail. Collared Sparrowhawks and Brown Goshawks breed in

Figure 3.4 Sugar Glider on Red Stringybark trunk.

these high altitude woodlands and Hobbies nest high in ravens' nests in tall trees and hunt birds, bats and invertebrates over the forest canopy. Southern Boobooks also prefer high and low altitude woodlands close to the city, and in the mountains. Though Wedge-tailed Eagles range through high mountain forest they tend to hunt and nest in cleared valleys. Little Eagles are not known to nest at higher elevations in this region.

Wet sclerophyll forest – These *montane zone* forests grow on the sheltered southern and east-facing slopes between 830 and 1500 m. Below 1100 m the forest is dominated by Brown Barrel *Eucalyptus fastigata*, Narrow-leafed Peppermint *E. radiata* and Manna Gum *E. viminalis*. Above 1100 m there is Alpine Ash *E. delegatensis* and Mountain Gum *E. dalrympleana*. Alpine Ashes are the tallest of these and dominant trees can grow to 40 m high with diameters of up to 1.5 m. The canopy is often closed. In the understorey, Blackwood *Acacia melanoxylon* can grow to 18 m high with a lower stratum of broad-leafed shrubs and ferns, such as the Soft Tree Fern *Dicksonia antarctica*, forming a rich habitat for birds (Figure

Figure 3.5 Don Fletcher climbing to a Wedge-tailed Eagle nest in a Scribbly Gum.

Figure 3.6 Soft Tree Fern in Brindabella Ranges (Michael Maconachie).

3.6). This is prime breeding habitat for Powerful Owls, Boobooks, Hobbies, Sparrowhawks and Brown Goshawks and is apparently suitable for Grey Goshawks and Masked Owls which do nest in Victoria and parts of New South Wales but not in the ACT so far. The upper elevation breeding limit for these species elsewhere is not clear. Peregrines breed on cliffs here below 1200 m and granite outreaches seem less suitable for them than is sandstone, though they do nest on ledges between granite boulders (Figure 3.7). When searching for their nests, we look for 'red rock' where a slab has fallen away from a cliff face. These often form ledges and protective overhangs for Peregrine nests.

Subalpine woodland – With increasing altitude Alpine Ash and other species disappear and in the *subalpine zone*, over 1600 to at least 1800 m, there is subalpine woodland of Snow Gum *Eucalyptus pauciflora* and Alpine Snow Gum *E. niphophila*. These are small trees, seldom more than 10 m in height and often can only attain 5 m. They have a short bole but heads are often windswept and distorted by incessant mountain winds and they don't seem to form hollows big enough for owls. Ground cover is made up of small shrubs and herbs. A German term for this habitat is *Krummholz*. Continual exposure to fierce, freezing winds causes vegetation to become stunted and deformed. Under these conditions, trees can only survive where they are sheltered by rock formations or snow cover. As the

Figure 3.7 Close to the 1200 m upper breeding limit, a Peregrine Falcon's nest between granite boulders in Namadgi National Park.

Figure 3.8 Snow Gum on Mt Gingera at 1857 m, Namadgi National Park, above the known breeding limit of any Australian raptor (Michael Maconachie).

lower portion of these trees continues to grow, the coverage becomes extremely dense near the ground (Figures 3.8, 3.9, 3.10, 3.11). These elevations are too high for breeding raptors (Table 2.1).

Alpine communities – Tree growth generally ceases at around 1800 to 2000 m in the Kosciuszko National Park but between 1700 and 1770 m in the more southern Victorian Alps. Above the tree line are typical alpine communities: tall alpine herb fields, feldmark, sod tussock grasslands, heaths, bogs and fens, and specialised bogs. Feldmark, also spelt *fjaeldmark* (from Norwegian, meaning 'mountain field'), is a plant community characteristic of sites where plant growth is severely restricted by extremes of cold and by exposure to wind, typical of alpine tundra and sub-Antarctic environments. Superficially these look much like those in high altitude North America and Eurasia. These Australian alpine communities are visited by Australasian Pipits *Anthus novaeseelandiae* and some other species wandering from lower levels during the summer, but there are insufficient resources there to support breeding raptors and the climate is too hostile for breeding. Winds are strong, soils thin, infertile and dry, and plants at these altitudes such as Mountain Pepper *Tasmannia lanceolata* have red pigment in their stems to absorb heat. A few Red-necked Wallabies *Macropus rufogriseus* and Common Wombats *Vombatus ursinus* live here, mainly in the subalpine zone, as well as nocturnal species such as the Mountain Pygmy Possum *Burramys parvus,* but no animals in numbers great enough to support breeding raptors.

Figure 3.9 Mt Franklin at 1646 m on the ACT–NSW border, above the known breeding limit of any Australian raptor (Michael Maconachie).

Figure 3.10 Mt Ginini (Michael Maconachie).

Figure 3.11 Snow Gum (Michael Maconachie).

4

Basics of raptor identification

Below are some basic points to help you with raptor identification. When identifying a raptor you should first ask yourself, *Is it a raptor?* Flying pelicans are as large as eagles, cormorants can resemble large falcons, Rock Doves can resemble small falcons. How do you differentiate them? The first step is to understand the basic raptor shape, and raptor size compared to some common high country birds (Figure 4.1).

Next, consider three characteristics of raptors important for species identification. First, male raptors tend to be smaller than females. Sometimes in flight this is not easy to detect because some males have proportionally longer wings than females. In perched Collared Sparrowhawks, Brown Goshawks, Peregrine or Brown Falcons, the size difference is more obvious, but less so in perched large and *Elanus* kites, eagles, kestrels or Hobbies. Second, juvenile raptors are often coloured differently from adults, for example adult Collared Sparrowhawks and Brown Goshawks are grey on the back and rufous-barred on the front (in spite of the confusing name 'Brown' Goshawk) whereas juveniles are brown, so can be confused with other brown raptors such as Brown Falcons, male Little Eagles or juvenile Peregrines and Hobbies. Juvenile Black-shouldered Kites have brownish breasts, heads and backs and can be confused with kestrels or other raptors. Third, some raptors come in different colour morphs, for example, light and dark-morph Little Eagles and maybe light and darker-morph Brown Falcons, though there is good evidence that these Brown Falcon 'morphs' are mostly

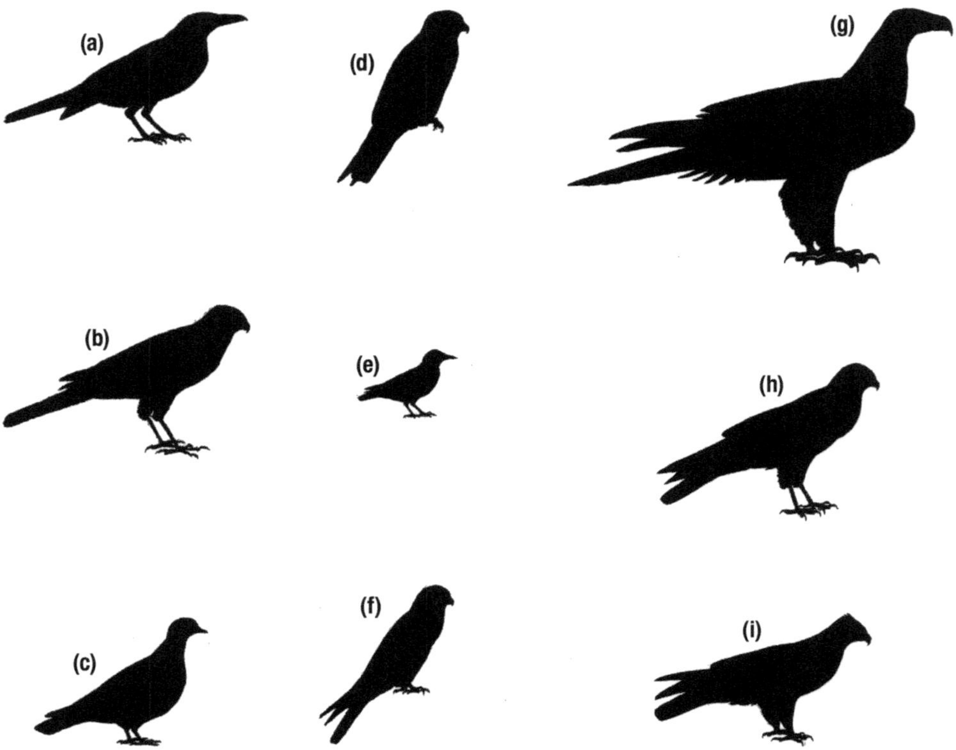

Figure 4.1 Basic shapes and sizes of: (a) Pied Currawong, (b) Collared Sparrowhawk, (c) Rock Dove (Feral Pigeon), (d) Black-shouldered Kite, (e) Common Starling, (f) Nankeen Kestrel, (g) Wedge-tailed Eagle, (h) Swamp Harrier and (i) Little Eagle.

different sexes at progressively different ages (McDonald 2003a,b). The key for overcoming these three obstacles is to learn the adult and juvenile plumages and alleged colour morphs.

Wind messes up raptor identification because wind changes a raptor's shape. A raptor such as a Peregrine that normally sits upright on a fence post will lean forward in the wind looking more like a corvid. A raptor flying with broad wings and splayed flight feathers, such as a Wedge-tailed Eagle, will tighten its wings into a falcon-shape, fly with quicker wing beats, show no splayed flight feathers and look much smaller than normal in high wind. The same applies to Brown Falcons, Little Eagles and Whistling Kites.

Thermals can be more helpful. We don't have the clear hawk migrations found in Eurasia or North America where raptors circle up in 'kettles' of hundreds of individuals, but our raptors do use thermals for hunting and display, especially during the breeding season, and several species can circle up in one thermal. In thermal flight you can scale the raptor's size, and goshawks and sparrowhawks will orbit in smaller circles than Brown Falcons or Little Eagles. Small raptors such as kestrels tend to attack medium-sized or large raptors such as goshawks or eagles, not the other way around, and medium-sized raptors also tend to attack large species. The large or medium-sized raptors often defend themselves by shrugging off the smaller one, side-slipping, or rolling and presenting their lethal talons.

Next determine the raptor's general type from the descriptions below. In bird taxonomy (classification), as with all living things, the *genus* is the first part of a taxonomic name, denoting a category designed to include one or more species of plant or animal presumably of common phylogenetic origin. The genus is always italicised and capitalised in the first half of a binomial scientific name, for example, *Aquila audax* for Wedge-tailed Eagle. In raptor identification, it is helpful to recognise different genera, as well as broader groups. Within genera, learn to identify perched and flying Collared Sparrowhawks versus Brown Goshawks, Australian Hobbies versus Peregrines, or Brown Falcons versus kestrels as these can be confusing (Figure 4.2).

Diurnal raptors

Diurnal, or daylight raptors in the ACT, fall into six broad groups:

Small (pointed-wingtips) Kites: These small, gull-like kites, the *Elanus* kites (Figures 4.3, 4.4), are smaller than a raven and represented in Australia by the Letter-winged Kite and Black-shouldered Kite. All *Elanus* kites look remarkably similar; they have short legs, long pointed wings with no visible 'fingered' ends to the primaries, and stubby toes for grasping rodents. These kites, including the Eurasian Black-winged Kite and North and South American White-tailed Kite are specialists of small rodents, especially the House Mouse *Mus musculus*. Unlike most other diurnal raptors, they seem geared to a 'boom and bust' (r-selected) strategy where they follow prey to different regions, breed up in large numbers, and breed more than once in a season or year, then disperse or die when prey disappears.

Tips: *Elanus* kites perch upright on exposed branches or poles, and hover over grassland. The magpie-sized adults appear wholly white from a distance, juveniles are washed with a brownish tinge. Perched, they appear legless, because of the short tarsi, with big heads and small beaks, and often they pump their tail up and down. Perched, and in flight, these are the most falcon-like hawks in Australia.

Large (rounded-wingtips) Kites: These kites, characterised by the genera *Milvus*, Black Kite, and *Haliastur*, Whistling (Figures 4.5, 4.6) and Brahminy Kites, are larger than a raven and characterised by easy, soaring flight with deeply 'fingered' primaries. They have short legs, and have supposedly weaker feet and beaks than hawks or eagles. Black Kites are cosmopolitan, maybe the most numerous raptors in the world, but they only occasionally occur in the Australian high country, and haven't been recorded as breeders. Whistling Kites are scarce breeders.

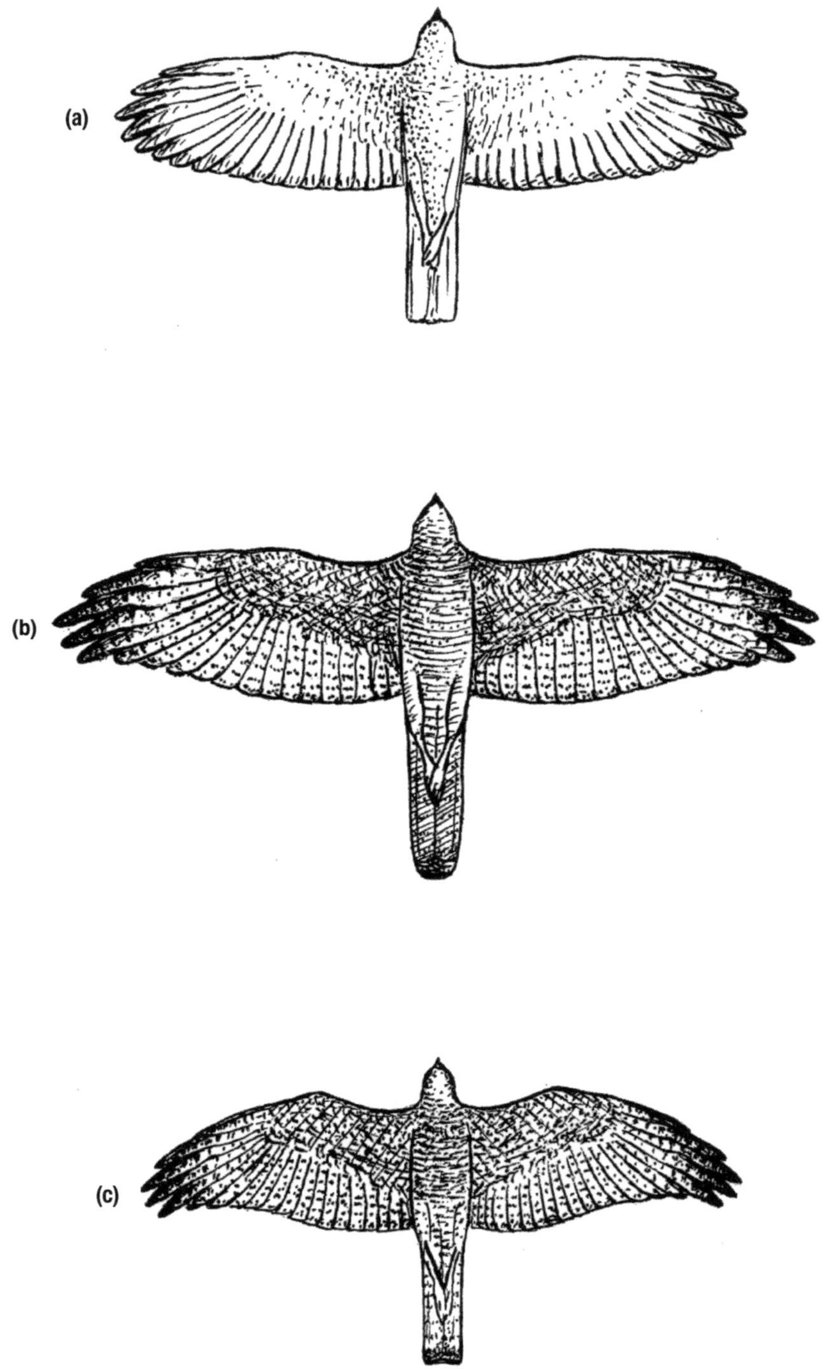

Figure 4.2 (a)–(c) Underwing views of high country raptors: (a) Grey Goshawk, (b) Brown Goshawk, (c) Collared Sparrowhawk.

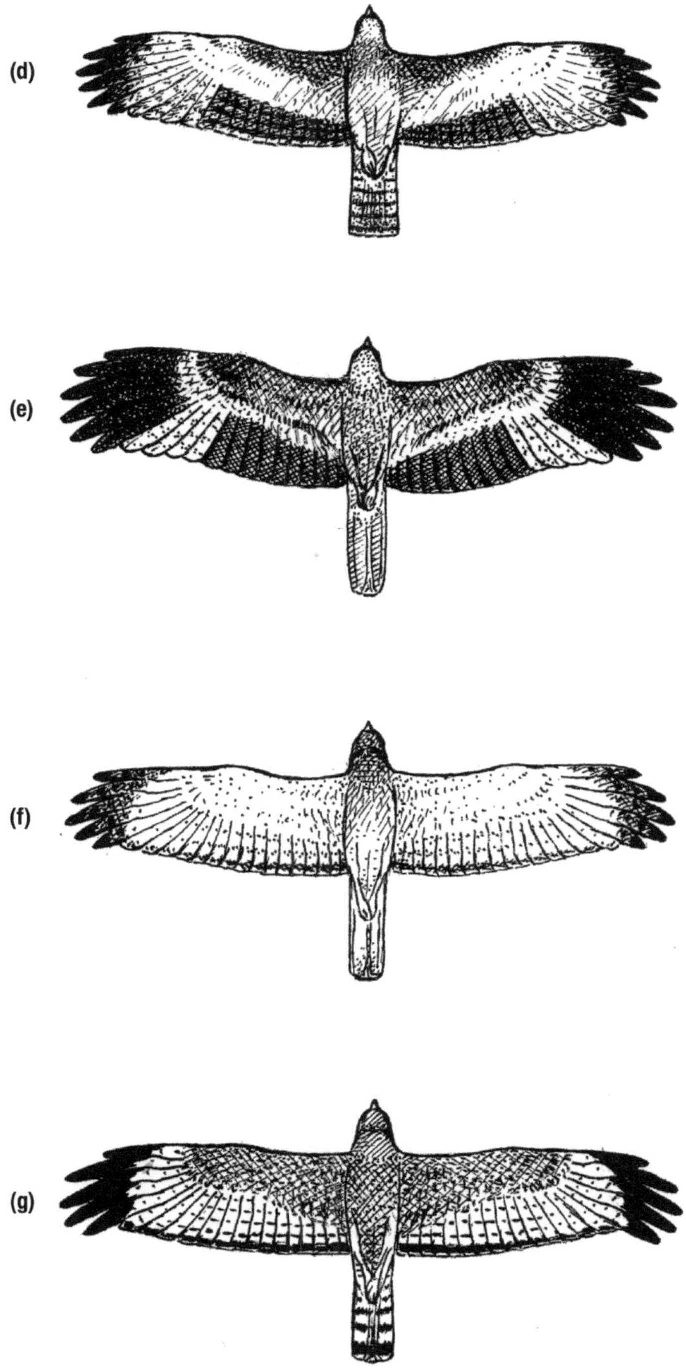

Figure 4.2 (d)–(g) Underwing views of high country raptors: (d) Little Eagle, (e) Whistling Kite, (f) Swamp Harrier, (g) Spotted Harrier.

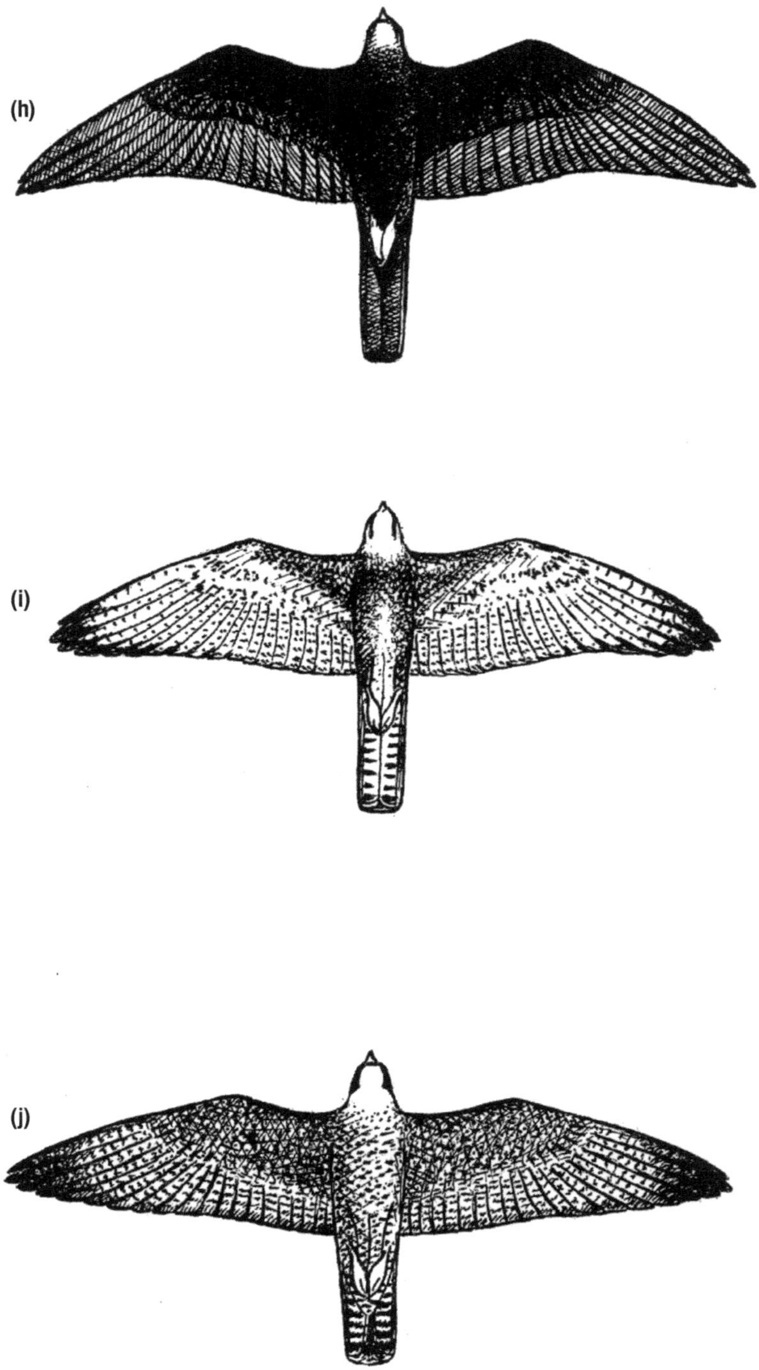

Figure 4.2 (h)–(j) Underwing views of high country raptors: (h) Black Falcon, (i) Brown Falcon, (j) Peregrine Falcon.

Figure 4.2 (k)–(m) Underwing views of high country raptors: (k) Wedge-tailed Eagle, (l) White-bellied Sea-Eagle, (m) Osprey.

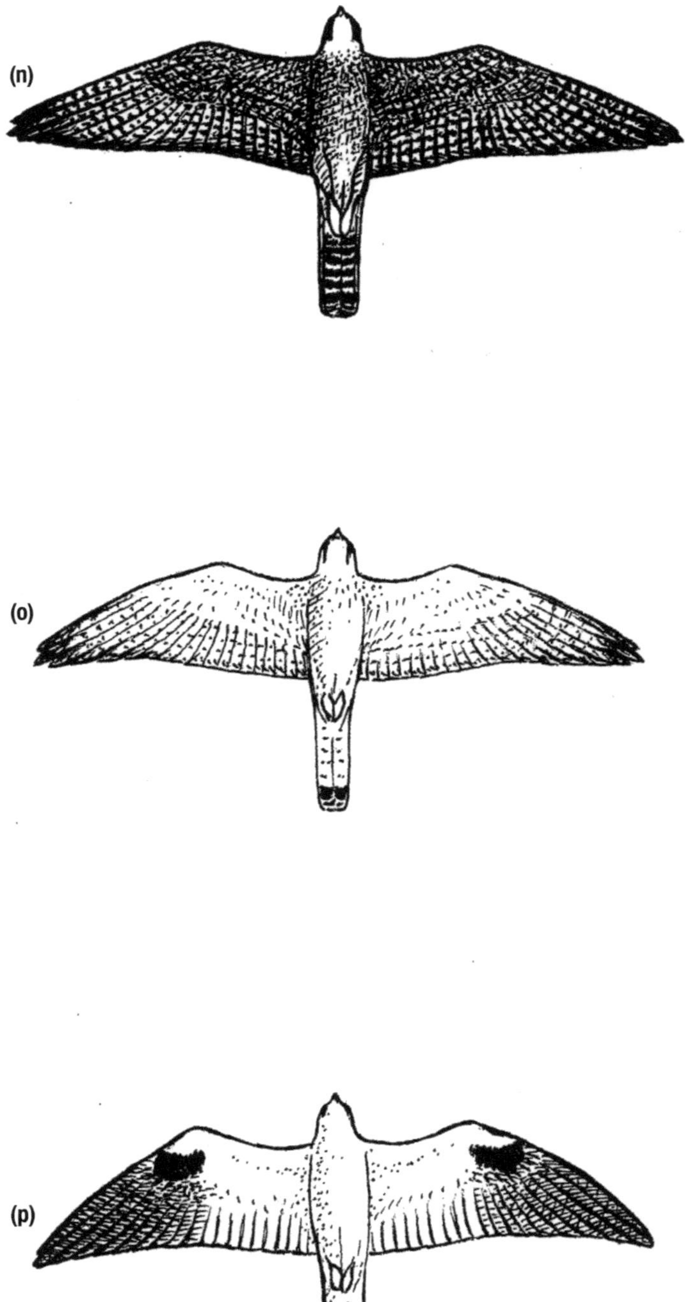

Figure 4.2 (n)–(p) Underwing views of high country raptors: (n) Australian Hobby, (o) Nankeen Kestrel and (p) Black-shouldered Kite.

Figure 4.3 Black-shouldered Kites have stubby toes for grasping rodents.

Figure 4.4 Black-shouldered Kites are small, gull-like kites (Julian Robinson).

Figure 4.5 Whistling Kite and Noisy Friarbird.

Figure 4.6 Whistling Kite with nesting material (Roger Williams).

Tips: Perched, the large kites look 'legless' because of the short tarsi. Little Eagles have longer legs, and harriers much longer legs. Look for the light sandy head on Whistling Kites, charcoal colour on Black Kites, deep dihedral in the wings and white patch on the rump of Swamp Harriers. In flying birds, get a clear view of the tail to separate the round-tailed Whistling Kite from the square-tailed Little Eagle or slightly fork-tailed Black Kite.

Eagles: The world's eagles fall into two groups: the 'true' or booted eagle, have feathered legs (feathered tarsi) down to their toes, as in the Wedge-tailed Eagle and Little Eagle (Figures 4.7, 4.8), both of which breed in the high country; and the other eagle group, 'sea-eagles', in the genus *Haliaeetus,* have bare legs (tarsi) and are more closely related to the *Milvus* and *Haliastur* kites, but as predators, the sea-eagles are just as powerful as 'true eagles'. These include the Australasian White-bellied Sea-Eagle (Figure 4.9), which occurs in the high country and breeds there.

Tips: Wedge-tailed Eagles and White-bellied Sea-Eagles are Australia's largest raptors. Watch for Sea-Eagles near fresh or salt water, Wedge-tailed Eagles usually away from towns, though not always. Perched, Sea-Eagles can look wholly white from a distance, Wedge-tails wholly black; look for wingtips extended beyond the tail-tip in Sea-Eagles, not so in Wedge-tailed Eagles which have an unusually long tail for a large eagle. In flight, look for a short tail in the Sea-Eagle, a long boat-shaped tail in the Wedge-tail. Much smaller Little Eagles are easily confused with medium-sized raptors such as Whistling Kites, Black Kites, Swamp Harriers, and even Brown Falcons. In perched Little Eagles, look for a stocky eagle; the light morph can look white from a distance, dark morph black. Feathered lower legs and a square tail discriminate Little Eagles from Whistling Kites, Black Kites, Swamp Harriers and Brown Falcons. In flight, Little Eagles soar with wings flat and a square-cut tail, but Whistling and Black Kites soar with wings drooped and rounded and notched tails respectively. Harriers usually quarter grassland closer to the ground, with wings in a characteristic dihedral.

Falcons: The cosmopolitan Peregrine Falcon (Figure 4.10) is considered by many to be the 'ultimate' falcon, with its muscled chest, pointed wings, extraordinary speed, and its ability to capture birds in midair. Most falcons are slow to accelerate, but they can sustain pursuit of prey, often chasing birds over long distances. Most raptors in the genus *Falco* have more-or-less pointed wings, though the Australian Brown Falcon's (Figure 4.11) are less so. Wings of most falcons have a low camber (flat profile) and can be swept back like a jet fighter. The trailing edge of the wing has a flare that joins it to the body, making a smooth line, unlike other raptors where the trailing edge dips into the body making more of a leaf shape. Brown Falcons break this rule, as if they are half-hawk, half-falcon. All falcons have a single tooth in the upper mandible used for cutting the spinal cord and dispatching prey, and a dark mark under their eye like a Cheetah. They have strong feet. All but the Black (Figure 4.12) and Grey Falcon breed in the high country, and Black Falcons occur as vagrants, though now rarely.

Figure 4.7 Wedge-tailed Eagle (Mark Osgood).

Figure 4.8 Little Eagle.

Figure 4.9 White-bellied Sea-Eagle (Julian Robinson).

Figure 4.10 Peregrine Falcon (Mark Osgood).

Figure 4.11 Underwing pattern of dark Brown Falcon.

Figure 4.12 Underwing pattern of Black Falcon.

Figure 4.13 Australian Hobbies (Julian Robinson).

Tips: Black-shouldered Kites have similarly pointed wings but high country falcons are generally darker, though kestrels and older Brown Falcons can appear white from a distance. Australian Hobbies (Figure 4.13) can be mistaken for Peregrines, or for kestrels when colours aren't evident. Most falcons have very similar calls, with diagnostic defence calls which are variations on *cac-cac-cac*, which can sound like a chitter in smaller species. The Brown Falcon is the outlier, with a chicken-like *kaaark* call, rounded wingtips, and wings held in a dihedral in flight. Other falcons hold their wings more or less flat.

In perched birds look for the diagnostic dark malar stripe on the cheek, double marked in Brown Falcons. In flight watch for pointed wings and rapid forward flight, though kestrels regularly hover, and Brown Falcons occasionally hover facing into the wind. Soaring Brown Falcons (Figure 4.14) can be confused with soaring Brown Goshawks. Look for the malar stripe on the Brown Falcon.

Figure 4.14 Brown Falcon (Geoff Dabb).

Accipiters: The short-winged, long-tailed woodland and forest hawks in the genus *Accipiter* occur worldwide in three forms: sparrowhawks have long toes for catching birds in midair, goshawks have shorter toes for capturing terrestrial mammals, reptiles and birds, as well as birds in midair, and some, like the American Cooper's Hawk, fall between the two classifications. Accipiters pursue prey with verve and agility in short sprints. They are quick to accelerate but they seldom sustain long chases. Australia has three accipiters, the Grey (White) Goshawk, Brown Goshawk and Collared Sparrowhawk (Figures 4.15 to 4.18), all of which occur in the high country, though only the Brown Goshawk and Collared Sparrowhawk regularly breed there.

Tips: Perched Brown Goshawks and Collared Sparrowhawks (Figures 4.19, 4.20), seldom far from cover, are small to medium-sized hawks that look leggy and small-headed with grey backs and rufous-barred fronts in adults; juveniles are brown. Grey Goshawks are greyish above and white below with some breast

Figure 4.15 Adult female Brown Falcon, juvenile male (White) Grey Goshawk and juvenile female Brown Goshawk. These birds were trapped in the ACT, used in a trial for reducing Silver Gulls at airports, then released.

barring; the white morph is wholly white. In the high country they almost never soar because they don't often breed here. The long tails, which all three sometimes twitch from side to side when perched, are diagnostic. In flight look for a square-cut tail-end on sparrowhawks and Grey Goshawks and a rounded tail-end on the

Figure 4.16 Juvenile male Black Falcon, adult female Brown Falcon, juvenile male (White) Grey Goshawk and juvenile female Brown Goshawk.

Figure 4.17 Juvenile Collared Sparrowhawk (Julian Robinson).

Figure 4.18 Juvenile Brown Goshawk.

Figure 4.19 Adult female Brown Goshawk.

Brown Goshawk. You will read in books that goshawks have short, rounded wings, but this is not true of the Brown Goshawk, which has longer, more pointed wings than most accipiters, so generic descriptions of this group do not hold for this species. However, accipiters do have less pointed wings than falcons. Another problem is that adult Brown Goshawks are grey, not brown, so birders commonly

Figure 4.20 Adult female Collared Sparrowhawk.

mistake them for Grey Goshawks. Grey Goshawks in the high country, as mentioned above, almost never soar, and they do not have rufous-barred or brown fronts. Sparrowhawks are the size of a Magpie-lark *Grallina cyanoleuca* (males) or Rock Dove (Feral Pigeon) *Columba livia* (females), Brown Goshawks are closer to crow-size and more easily confused with Brown Falcons, but have smaller heads.

Harriers: Hawks in the 'circling' genus, *Circus*, occur worldwide, and mostly nest on the ground, except for the Spotted Harrier. They have acute hearing and a

Figure 4.21 Swamp Harrier (Duade Paton).

facial disc like an owl, which apparently augments and funnels sound. In flight they quarter low and slowly over grassland on long blunt-ended wings with fingered primaries. Their long legs help them snatch prey in tangled vegetation. Australia has two species, the Spotted Harrier and Swamp Harrier (Figure 4.21), both of which occur in the high country, but only the Swamp Harrier breeds here and, at the time of writing, there are only one or two pairs in the ACT. Swamp Harriers can be polygamous, one breeding male with two or more females in different nests.

Tips: Slender body, owlish face and facial ruff, long legs and long wings with squarish ends, and their habit of standing on the ground all characterise harriers.

They quarter slowly over grassland and swamp, often only a few metres up, but they also soar on long wings. They appear dark, and the white rump patch on the Swamp Harrier separates them from Little Eagles, Brown Falcons and large kites.

Nocturnal raptors

The world's owls, the Strigiformes, are divided into two families: 1) Tytonidae – 'bay owls' or 'barn owls'; and 2) Strigidae – 'typical owls'. The two families show similarities in behaviour, breeding and methods of hunting; owls are similar around the world in behaviour and appearance. Barn owls are said to be distinguished from 'typical' owls by a sharply defined facial disc, typically heart-shaped (Higgins 1999) but, to my eye, the differences are not always that clear. For instance some individual Southern Boobooks have a fairly well-defined facial disc. In Australia the two groups are represented by the genera *Tyto* or *Ninox*.

Tyto owls can be divided into four groups: Sooty Owls, Masked Owls, Barn Owls and Grass Owls (Figures 4.22, 4.23) and all four are represented in Australia,

Figure 4.22 Barn Owl.

Figure 4.23 Masked Owl.

the Barn and Grass Owls preferring open land, and the Sooty and Masked Owls preferring forested land. No *Tyto* owls have been found breeding in the ACT, but they do breed in the high country and the Barn Owl is a regular visitor.

Ninox in Australia is represented by five species, depending on taxonomic treatment. Unlike some *Tyto* species, all breeding *Ninox* prefer forest or woodland to open grassland. In the high country only the Southern Boobook is common. Rufous Owls are never seen, Powerful Owls (Figures 4.24, 4.25) are rare breeders in the mountains, and Barking Owls are rare visitors.

All *Tyto* have a characteristic head-swaying threat display that no *Ninox* have. *Tyto* owls tend to have a variety of screeching calls but all Australian *Ninox* have a characteristic two-note *hoot* – higher pitched in Boobooks, lower pitched in Powerful Owls. In the ACT you are only likely to see three owl species: Barn Owl, Southern Boobook, and rarely, the Powerful Owl. But the Tawny Frogmouth, not an owl at all, is the most commonly seen nocturnal bird, slightly larger than a

Figure 4.24 Southern Boobook after robbing a nest of Common Mynas.

Boobook and without the compressed beak, and 'killer' feet and talons that characterise raptors.

Tips: Boobooks are magpie-sized brownish owls, slightly smaller than a Tawny Frogmouth in dimensions, but heavier (frogmouths are mostly feathers). In the high country you are more likely to hear the two birds than see them. The Boobooks' characteristic two-note *boo-book* call is something like the call put in a European cuckoo clock, and differs from the Tawny Frogmouth's monotonous one-note *oom oom oom oom oom*. Barn Owls appear white and are most often found at day roosts, or seen flying. Seldom heard, they make a loud screech when they do call, as do Masked Owls. Sooty Owls give a long descending siren-like scream or wail, known as the 'falling-bomb' call, and harsh screams similar to a Masked Owl.

Figure 4.25 Powerful Owl.

In Canberra, many owls are harassed and disabled by magpies and currawongs, so listen for mobbing Noisy Miners, friarbirds, currawongs, magpies, cockatoos and other birds that can lead you to roosting Boobooks or Barn Owls.

5

Watching raptors

It is against the law to disturb nesting raptors. Generally, raptors don't like people or other predators to be higher than they are, looking down on them, so it is important to stay below the nest and avoid causing distress. Disturbance by ornithologists and birdwatchers raises a vexed question. When ornithologists and authorities are told about a disturbance-prone nest, the location becomes widely known, more people visit the nest, and the pair is further disturbed. The best solution is education about disturbance and requests for care and cooperation, but precise nest locations of sensitive pairs should remain secret.

To minimise disruption to raptors being watched or photographed, the following guidelines can be followed when visiting a nest site:

- Visit the nest site only as often as necessary.
- Visit the nest with sensible regard to weather, e.g. not during rain or on very cold nights when nestlings can be chilled.
- Where possible, visit the nest in the latter half of the nestling period; make your visits short, 30 minutes or under, unless prolonged observation is necessary. Take special care during the first week after hatching.
- Raptors are as variable as people; use common sense, be sensitive to their behaviour.
- Read behaviour. A raptor standing with one foot raised partly hidden by belly feathers and throat feathers puffed out, or preening, or raising its plumage, is relaxed. It is best if the raptor doesn't watch the observer and keeps feeding or

whatever it was doing, that is, it does not stop and watch you. If it lowers its foot and stands on two feet, sleeks its plumage, defecates, adopts a horizontal body posture, stretches its neck or bobs its head, it may be startled and about to leave.

- Where possible, approach the nest from the front, below and in clear view of the birds (as mentioned before, raptors dislike humans being above them), talking as you approach so the raptors hear you coming and are not surprised.
- For behavioural studies close to the nest: (a) be predictable, relaxed and avoid sudden movement or noises; (b) gradually habituate birds to your presence; (c) choose birds whose breeding success is known; and (d) watch from a distance with a telescope, using a hide if necessary.
- Do prolonged observation sessions only after the raptors are habituated to you and will forage and feed young in your presence. We find that certain female Boobooks watch us and refuse to hunt or feed in our presence, so we leave them alone.
- Avoid disturbing the local habitat. Do not remove any vegetation near the nest.

It is popular to use call playback to lure in owls for birdwatching and photography. In the Corkscrew Swamp in Florida birdwatchers using playback ruined the breeding attempts of every breeding pair of Barred Owls in one year (de la Torre 1990). Disturbance by birdwatchers could have caused the failures of Powerful Owls at Lee's Creek in the Brindabella Ranges. Some *Ninox* differ from this, and they tend to eventually ignore broadcasts.

6
Finding raptors

We struggled over a wooded hill on the Murrumbidgee River. The afternoon was hot now without any wind, and David Mallinson stopped to look through binoculars.

'I remember it as being just over there', he said.

We glassed the opposite hillside but couldn't see anything.

Some years later, close to where we were walking, David discovered a new plant species, the Tuggeranong Lignum *Muehlenbeckia tuggeranong* known only from flood terraces on the eastern bank of the Murrumbidgee River. But today we were searching for eagles. Rocks crunched underfoot and we pushed branches aside as we searched for the signs – eagle feathers, stripes of whitewash on the ground, pieces of rabbit or kangaroo, and listened for mobbing currawongs or ravens, the enemies of eagles. This Wedge-tailed Eagle nest was one of the few left around Canberra in 1985 since the construction of houses and walking trails had disturbed many pairs. Leopold and Wolfe (1970) had mapped the nests they found in Canberra, on Gungahlin Hill, Mt Wanniassa, but these nests were all abandoned by 1985. The year before, this pair had two young, but we stood under the old nest on the Canberra bank of the river and could see no whitewash under the tree. Because eagles often build several nests in their home range and switch nests from year to year, there probably were other nests across the river, or maybe some on this side hidden by forest. We pushed through tea-tree, solid as a thicket, under branches and between trees. As we struggled up a steep ridge we started to wonder if the eagles had taken a year off.

We turned back, followed the ridge north and stumbled down a long graded hill, with the binoculars swinging around our necks, into an open area along the riverbank, a beach. It was full of naked people. The nude people watched us, with our binoculars swinging. 'Oh, great', said David, as we pushed ahead, studying the toes of our boots, stupidly hoping that no-one would notice the 'perverts', passing reclining nudes and people talking in groups.

We didn't find any Wedge-tailed Eagles nesting near Canberra that year, but a few years later, David found a pair near his house that had recolonised an old Leopold and Wolfe territory, displacing some Little Eagles. They probably weren't the same pair, those had been gone for 5 years or so, but this pair was more accustomed to humans, and later a few more started nesting closer to houses.

Though the Australian high country has fairly good populations of breeding and transient raptors, it can be hard to find them. It's not quite as good a place for randomly spotting perched or flying raptors as, say, western New South Wales or even the built-up areas around San Francisco or Marin County in California. More open areas away from forest such as Burrinjuck Dam, Wee Jasper and the Monaro Plain can be good in late summer for spotting raptors. The difficulty here is partly due to the topography and vegetation, and the way certain raptors use trees to perch in, hunt from, and hide in. Because of the hilly and mountainous high country, we have fewer wide vistas to view raptors at a distance and less flat, heated ground to generate thermal activity for soaring. However, you can see Wedge-tailed Eagles, Peregrines and Little Eagles using diverted sheets of air bounced upward along the Lake George escarpment, along with wary hang-gliders dodging attacking territorial eagles (Figure 6.1).

Juvenile and immature Wedge-tailed Eagles cluster together in summer and autumn, especially along rivers and in places like Wee Jasper. Brown Falcons, Black Falcons, kestrels and other species are visible on the Monaro Plain in summer. Much of the region is forested and raptors can be difficult to see.

The numbers of visible diurnal raptors in the Australian high country are low compared with those at similar latitudes in places like southern Spain, Arizona or California where you see dozens of raptors circling in the sky during even a short bus ride. These reduced detection rates are tied to particular species. In south-eastern Australia we have few obvious large kites – Black Kites and Whistling Kites – compared with northern Australia. Australia has no vultures, and no ubiquitous soaring or roadside *Buteos* such as the North American Red-tailed Hawk, and no major raptor migrations such as those you see in southern Spain, Hawk Mountain in Pennsylvania, or the Golden Gate Park near San Francisco, places where you can see hundreds of migrating raptors in a day. Australian raptors have no Africa or South America to migrate to, so our Black Kites are nomadic or relatively sedentary, whereas European Black Kites mostly leave, then funnel back north, through places like Gibraltar and are highly visible. Our Little

Figure 6.1 Wedge-tailed Eagle attacking hang glider at Lake George, New South Wales (Andrew Luton).

Eagles move from their breeding territories in winter but stay in the area, as far as we know, so some are visible. Closely related Booted Eagles mostly leave Europe, even leave warm southern Spain for Africa, then return in spring. In northern North America, Golden Eagles and Peregrines migrate south after breeding, but Australian Wedge-tailed Eagles and Peregrines stay on their breeding territory over winter, so you look for them at their breeding site, not on migration routes. High country Brown Goshawks, Collared Sparrowhawks, Hobbies, kestrels and Swamp Harriers do leave their breeding territory for the winter and seem to migrate, maybe, to northern Australia, New Guinea and Wallacea. Torres Strait is one of the pathways for birds migrating north and there is migration across Bass Strait for Swamp Harriers.

During the breeding season our breeding Brown Goshawks are extremely shy compared to breeding Cooper's Hawks which nest in city front yards and hunt, feed and mate in plain view. Our large and small eagles are more timid than similar-sized North American *Buteos* or the Bald Eagle, though Little Eagles are becoming tamer and now nest closer to human habitation.

The high country is rich in raptors if you know where to look and how to look for them, understand how to watch and listen for signs, the habitat they use, and how and when they hunt. North American accipiters such as Cooper's Hawks and Northern Goshawks 'sing' at dawn and this is a valuable tool for locating them. However, this habit has not been confirmed in Australian accipiters, though Brown Goshawks, especially females and young, do call from their nest using a single note not unlike the call of a Sacred Kingfisher; this can help observers locate nests.

Ornithologists in the United States use playback to discover if accipiters are in woodland, much as Australian researchers use playback to locate owls. However, playback was tested with Brown Goshawks in the ACT and was only partially successful. Only a few females responded.

Watch for mobbing or alarmed passerines, or birds such as ducks that freeze and look skywards. Follow their gaze and scan with binoculars. White-plumed Honeyeaters are particularly good at alerting you to raptors; look up and search the sky any time you hear their alarm call. Other passerines can help, for example perched friarbirds and Common Starlings. The starlings give a rapid-fire *chink-chink-chink* call when an avian predator is nearby so always look up when you hear this, even though starlings may be calling about non-raptors such as a currawong. Stop, scan the sky with binoculars, wait and check the horizon for a raptor to appear. Scolding Eurasian Blackbirds and Red Wattlebirds often reveal goshawks and sparrowhawks, and currawongs and magpies are particularly aggressive towards raptors, especially owls and Little Eagles. Though species such as Grey Fantails, White-throated Treecreepers and Noisy Miners also mob raptors, they call so often about non-raptors and each other that they're of little use for reliably spotting raptors unless you can learn to discriminate the different calls they use for falcons, eagles, accipiters or non-raptors. However, Noisy Miners will often perch on the lip of a hollow and call as an owl is emerging, revealing its nest. Flying Common Starlings are also helpful for spotting raptors; they often swarm in twisting flocks that rise and drop in the sky and change shape when an accipiter or Little Eagle is in the air, but dive into bushes when a Peregrine or Hobby appears.

Raptors attack and mob each other and can help you find other raptors and identify the raptor species through a sense of scale; small raptors generally attack and dive at larger ones (Figure 6.2). For example, kestrels attack goshawks, which attack Little Eagles, and they all attack Wedge-tailed Eagles. The mobbing bird stays above and behind the larger raptor, and the raptor may sometimes roll and present its feet in defence. Hobbies, Peregrines, kestrels and especially Brown Falcons are noisy when attacking enemies so it is important to learn their calls (described in the species accounts in Appendix 1). Little Eagles often call when flying high and Whistling Kites are the noisiest diurnal raptor in the high country. Both can be located by their calls and they often call close to nests. Female Brown Goshawks and Collared Sparrowhawks chatter at enemies, and adult females and their young call for food from the nest and nest area. Grey Goshawks make a loud single note, like a whistle, that helps locate them. Fledged young accipiters and falcons chase and tease each other and call loudly while doing so. Follow these calls and investigate.

A raptor carrying prey, greenery or sticks is often headed for a nest. If you plot where the raptor is headed and search that area carefully, you can often find it. Another tip is to look for 'butcher's blocks' or plucking logs where hawks or eagles pluck bird feathers. These are scattered in patches on the ground, often on a log or

Figure 6.2 Brown Goshawk attacking a Wedge-tailed Eagle (Mark Osgood).

perch just uphill from the nest. With eagles this is often a lateral branch in a tree, but Wedge-tailed Eagles and Brown Goshawks often use fallen logs if the log stands off the ground on struts of broken branches. You can also find a ring of feathers on the ground in forest and open land where a raptor has killed and eaten prey (different from a 'butcher's block' close to a nest). Feathers at these sites are plucked out, not chewed as a fox might do, and sometimes the ground is marked with telling stripes of whitewash. Places where a Little Eagle kills a Boobook, or a goshawk kills a Boobook or Hobby, can be scenes of great violence (Figure 6.3). These reveal where a raptor hunts, not necessarily where a raptor nests; raptors rarely hunt vertebrate prey from their nest.

A raptor's droppings are bright white, so white patches or streaking on a cliff can reveal the roost of a Peregrine (Figure 6.4) or kestrel. Wedge-tailed Eagles often roost on a rock and leave droppings, but more often they roost in a tree. We have it easier in Australia than in other parts of the world because only Peregrines and kestrels nest on cliffs in inland Australia, but many eagle, falcon, vulture, owl and other bird species nest on cliff ledges in other parts of the world (see Chapter 2). Whitewash splashed on the forest or woodland floor, or under a stick nest in a tree, can mean a raptor is using that locale. As in all birds, droppings come from a single source, the cloaca. The white part of the dropping is mostly composed of uric acid

Figure 6.3 Remains of a juvenile Southern Boobook killed by a neighbouring Brown Goshawk.

(from the kidneys), white because it is a crystal, like table salt. The dark portion (usually black, but green if the raptor's stomach is empty) is waste from the bowel. With practice, you can differentiate some raptor droppings from some other bird droppings, e.g. corvid droppings. Excrement of falcons and owls drops straight down from the raptor and often has a circular black dot in the middle of the white, but hawks and eagles tend to squirt the faeces and these can appear as short or long white lines or stripes under the perch or nest. Obviously, this depends on how high up the nest or perch is, and the dropping will be fragmented if dropped from a great height or through branches.

Raptors start moulting about the time females lay eggs so moulted flight and tail feathers on the ground can be close to a nest and these help the observer find secretive species such as Brown Goshawks or even Wedge-tailed Eagles (Figure 6.5).

In the end, searching for raptors takes vigilance, a harmony with nature and alertness to signalling birds; searching for nests mostly comes down to hard, systematic footwork with binoculars, checking likely nests and habitats, walking river and creek systems with big trees, and listening for tell-tale mobbing birds or calling raptors. A survey of eagle or hawk nests can take many weeks, even months of such walking. Nests can be surprisingly cryptic, even from aircraft (Figure 6.6).

Figure 6.4 Peregrine Falcon nest cliff on the Murrumbidgee River (Michael Maconachie).

Figure 6.5 Raptors moult during the breeding season and leave primary feathers close to the nest, like this Wedge-tailed Eagle feather.

Figure 6.6 Looking straight down from a helicopter onto a Wedge-tailed Eagle nest containing a single young.

7

Designs for survival

Being hit in the head by an attacking Peregrine is like being hit with a hammer. If you have no helmet, you see stars; the blow draws blood. It gives you sympathy for Galahs, pigeons or ducks killed by a stooping Peregrine. Some of the females are so lethal that you develop strategies to avoid being beaten up while you're collecting prey remains and bagging the young to haul them up a cliff for banding. Peregrines are less agile than goshawks or Brown Falcons, so you can place your back against the cliff wall, shelter under some overhanging rock or behind a lateral slab, and stick your boot out from cover to distract the attacker (Figures 7.1 to 7.3).

How can such a powerhouse, a warlike killing machine, be packed into such a bullet-shaped, compact body? Certain body designs and physiology give the Peregrine enormous striking power, and give the Brown Goshawk the agility to catch prey in heavy cover or whack you at its nest even if you try to take cover (they are too agile to evade), though not with the force of a Peregrine. In contrast to Peregrines, eagles and large kites have the ability to soar for hours using very little energy. Sparrowhawks can crash through undergrowth or slip through small gaps after prey. Hobbies can catch insects in midair, hold them in their fist like an ice-cream, and consume them without alighting. Brown Falcons can run down a snake and kill it. Below we examine some of the physical designs for survival that separate raptor species and give them special aptitudes for hunting, dealing with enemies, and breeding; these are drawn from several sources including Welty (1982), Grossman and Hamlet (1964), Campbell and Lack (1985), Johnsgard (1990, 2002), Ferguson-Lees and Christie (2001), Cartron (2010), Peeters and Peeters

Figure 7.1 Attacking Peregrine (Mark Osgood).

(2005), Peeters (2007), Duncan (2003), Burton (1973), Schodde and Mason (1980), König and Weick (2008), Marks *et al.* (1999), Debus (2009), Debus (2012), Marchant and Higgins (1993), Higgins (1999), Perrins (2009) and Olsen (2011a).

Diurnal raptors

Skeleton

A raptor's skeleton is characterised by two qualities – strength and weight reduction. Internal organs are crammed into a box-like, rigid airframe (have a look at a chicken carcass after Christmas dinner). Inflexibility in this airframe is necessary for the lifestyle that raptors lead, so the vertebral bones of the back are fused to the pelvis, except for a single vertebra that connects the two regions (Figure 7.4). To maintain strength and weight reduction, the long bones in a bird's skeleton, such as the wing bones, are supported by fine internal struts (Figure 7.5).

Figure 7.2 Distracting an attacking Peregrine with outstretched boot (Mark Osgood).

Figure 7.3 Typical injuries inflicted by attacking Peregrines.

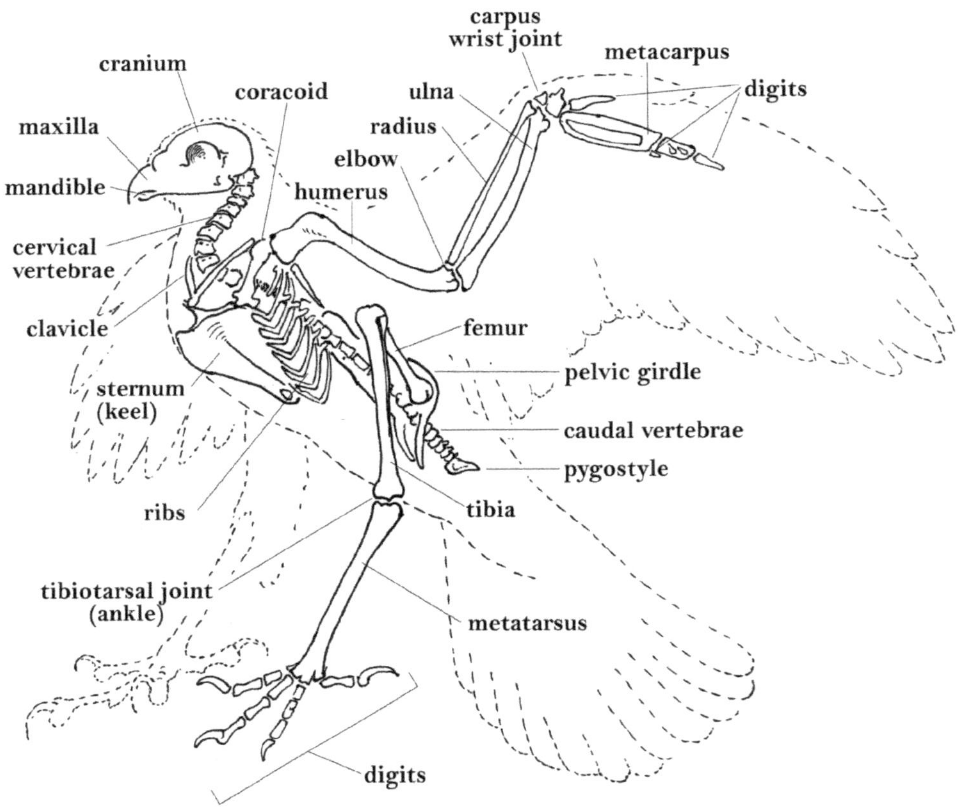

Figure 7.4 Skeleton of diurnal raptor.

Respiration

As in mammals, the respiratory system in birds delivers oxygen from the air outside through the trachea to the bird's tissues, and removes carbon dioxide and delivers this back through the trachea and outside the body. In addition, the avian respiratory system plays a crucial role in thermoregulation, maintaining normal body temperature. Birds need a super-efficient respiratory system because they

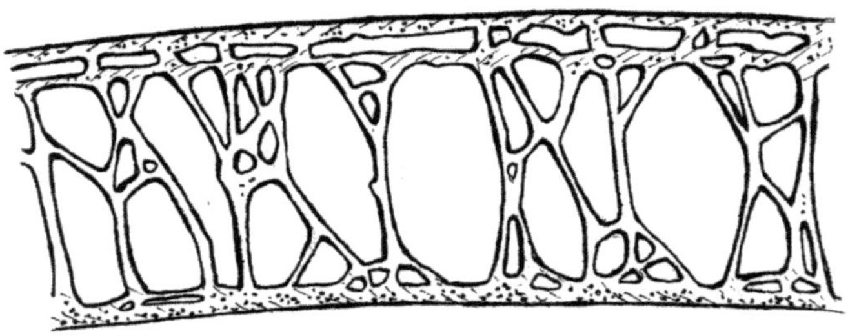

Figure 7.5 Internal struts in a bird bone.

have such high metabolic rates; they need to sustain the enormous work output demanded of flight, and flight generates enormous heat. Consequently, the avian respiratory system is different from that of other vertebrates; compared to other animals, birds have relatively small lungs plus eight or nine air sacs that are crucial to respiration (Figure 7.6) but not directly involved in the exchange of gases; these are found in the body cavity, inside muscles, and even reach into the wing bones and skull sinuses. Air sacs may also eliminate the heat generated by muscle friction.

Air sacs and the avian flow-through lung

Powered flight is energy- and oxygen-expensive, so the air sacs act as reservoirs and bellows that feed air into the raptor's lungs (lungs that are at least 10 times more efficient than a mammal's lungs). All tetrapods (mammals, reptiles, amphibians) except birds have a pair of lungs that operate on the bellows principle, like blowing air into a pair of balloons. In humans this is associated with respiratory movements of the diaphragm and rib cage. So, in humans, the lungs inflate and suck in air, then recoil and blow the air out again. That is, mammalian lungs inhale and exhale air by volume changes. This volume is increased by drawing in oxygen-rich air and decreased by expelling air that is partially depleted of oxygen and carries waste carbon dioxide to the outside. But the human lung never completely expels all the stale air. In contrast, a bird's lung does. Birds have a one-way flow of air through

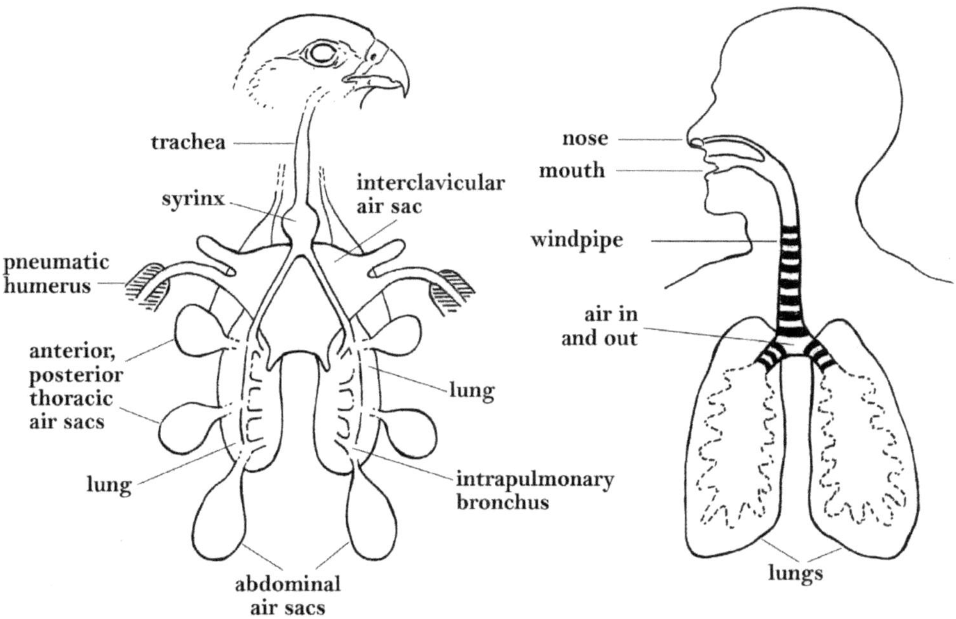

Figure 7.6 Avian flow-through lung of a raptor compared to the respiratory system in a mammal.

their lungs, not two-way. Bird lungs don't have the tiny air sacs (alveoli) that we have. Instead, they have tiny tubes called air capillaries; the air flows through these capillaries one-way so birds always have fresh air flowing through the lung, maintaining very high concentrations of oxygen up against the avian bloodstream.

The air sacs take up considerable space in the body cavity of the bird and they are classified as anterior (forward) and posterior (rear) air sacs, and are connected to the lungs so that, when a bird breathes air in, the air flows in through the bird's beak into the trachea and into the posterior air sac. The air then flows though the lungs in a single direction into the anterior air sacs through the fine set of air capillaries (the parabronchi) inside the lung.

The anterior air sacs also increase in volume – they pull air in through the lung capillaries. Then when the bird exhales it squeezes on these air sacs. The posterior air sac empties into the capillaries and the anterior air sac empties out into the bird's trachea, then out through the nostrils and beak. In this way, they always have air travelling one direction through the lung and, because it's always fresh air, the bird has very high oxygen content through the lung tissue. Because these air sacs also use spaces inside some of the bird's bones, the skeleton is highly pneumatised, including the shoulder bones and vertebrae. The bird breathes through the pneumatic humerus too (see Figure 7.4), and the bones are extremely light, an additional adaptation for flight.

The air capillaries in the lungs are closely surrounded by blood capillaries and this is where the exchange of oxygen and carbon dioxide takes place. Because of the arrangement of air sacs and lungs, air flows in the same direction through the lungs whether the bird is inhaling or exhaling (uni-directional flow-through). Unlike in other tetrapods, the avian lungs retain the same volume during breathing – the air is pumped by changes in the volume of the air sacs not the lungs.

Ornithologists used to think that pumping wings pumped air into the respiratory system but this is not so; only in crows and pigeons does a 1:1 synchrony occur.

Digestion

Compared to plant-eaters, the digestion system of a raptor is simple and short. Below the crop, really an empty sac, the gizzard grinds food but is not as muscular as, say, a chicken gizzard because it is grinding meat protein instead of grain. As with most other birds, weight reduction is crucial, though diurnal raptors commonly ingest stones as other birds do. The major components of the avian digestive system are the oral cavity, pharynx, oesophagus, which includes the crop in diurnal raptors but not owls, a stomach with two chambers – the proventriculus which adds hydrochloric acid and digestive enzymes to the food, and the gizzard (ventriculus) where food is mixed with these digestive agents, and where a part-digested bird or mouse is separated from bones, feathers or fur to produce pellets.

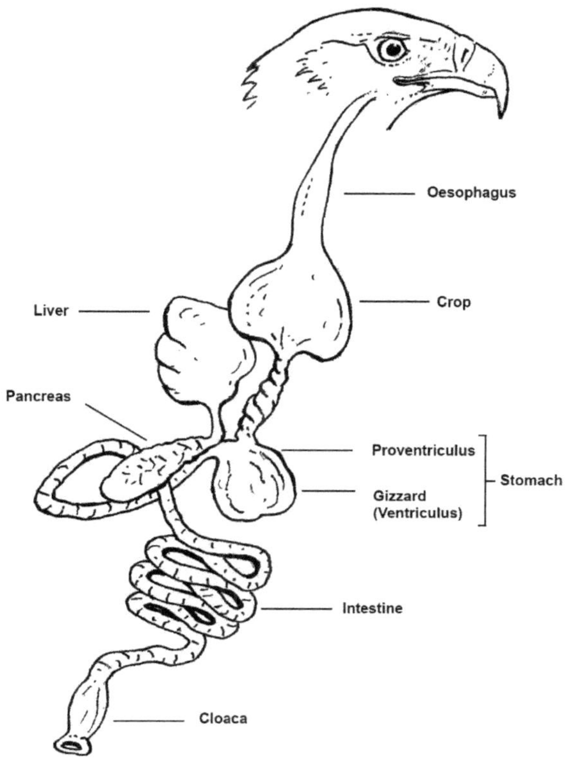

Figure 7.7 Digestive system of a raptor.

You can watch a diurnal raptor work this process with lateral, jerky movements of its head and neck as it pushes pieces of food down from the crop into the stomach. Below the gizzard are the small intestine, and the large intestine that empties into the cloaca, Latin for 'sewer' (Figure 7.7).

Feathers

The bodies and wings of birds have different types of feathers (Figure 7.8), and some believe that the contour feather is the fundamental feather type. Contour feathers consist of a shaft and side barbules that zip together. You can see this arrangement in a flight or tail feather, both types of contour feathers. Smaller overlapping contour feathers shape the bird's body, make it bullet-shaped and aerodynamic, and protect and insulate the body. Small muscles raise and lower these feathers to create varying thicknesses of insulation to heat the bird. Special contour feathers, the coverts, arch over the leading edge of the upper wing to create an aerofoil.

The tips of some flight feathers are deeply emarginated, as if half or more of the width of the primary has been snipped away leaving narrow tips, like fingers. These fingers are 'miniwings' and essentially extend the bird's wing and provide

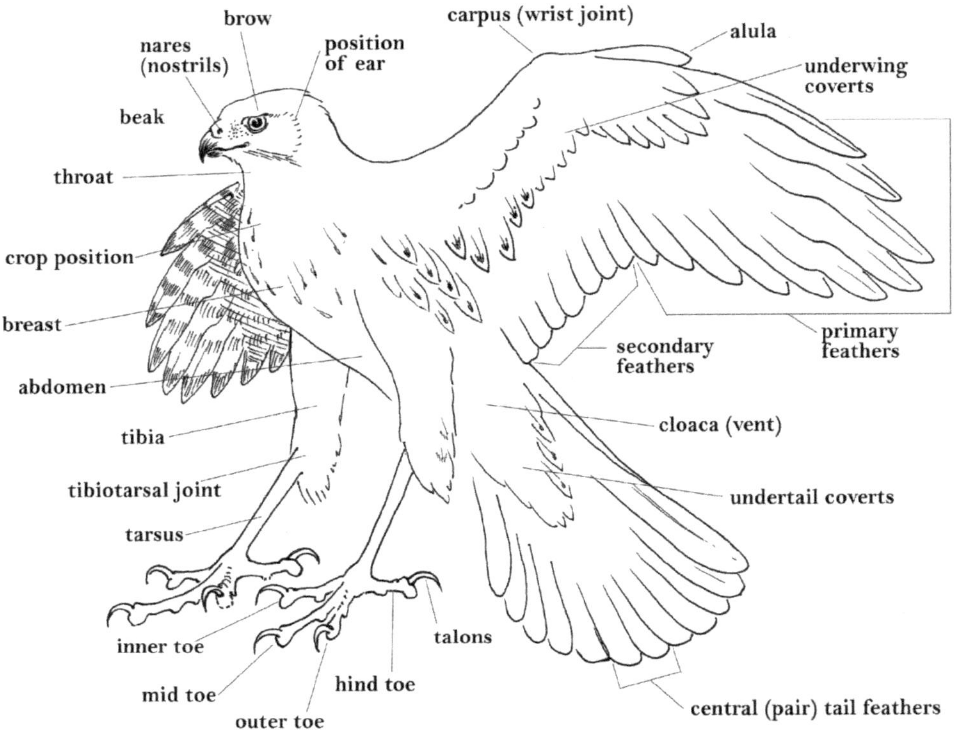

Figure 7.8 Body parts and feathers of a diurnal raptor.

extra lift (Tucker 1993). Brown Goshawks and Collared Sparrowhawks have extremely flexible primary tips for dashing through dense vegetation.

In flapping flight a raptor creates its own wind and uses its flexible wingtips to climb these air columns. The long outer flight feathers act as propellers, and in soaring flight you can see the spread, slotted primaries of an eagle bending and twisting independently of each other, to minimise air disturbance and increase lift. As mentioned above, flapping flight is energetically expensive and feather design helps to save energy. However, by developing such soaring wings, eagles, large kites and other raptors sacrifice speed. Falcons soar less than eagles, and falcons have far less wing slotting. Instead, they have great speed, especially in a stoop.

Another type of feather, the powder down feather, crumbles away at the tip and provides waterproofing. You can see this on the back of a goshawk, Peregrine or eagle as a shiny, greyish bloom (Olsen 1990). A black velvet cloth wiped over a Peregrine's back comes away grey (Peeters and Peeters 2005).

Feathers are resilient, but they are dead structures, and they abrade or can be broken if not cared for. Birds spend a lot of time grooming, taking oil from the uropygial gland at the base of the tail and running each wing and tail feather through the beak (Figure 7.9).

Figure 7.9 Southern Boobook spreading oil from its uropygial gland through its wing feathers.

Raptors bathe regularly. Water softens feathers and allows them to re-form their shape, with the raptor rousing and raising each feather so it falls into place; they preen each wet feather to clean and re-form it. Some raptors 'dust bathe' which may kill ectoparasites and also reduces extra oil and greasiness in the feathers that reduce feather performance.

In raptors, juvenile plumage is often very different from adult plumage, mostly the young ones appear browner in colour. Juvenile plumage is said to suppress aggression from adults. Mid-plumage, for example the second- or third-year plumage of a White-bellied Sea-Eagle, could generate more status with conspecifics than does first-year plumage. Juvenile tail and flight feathers are often, but not always, slightly longer than adult tail and flight feathers. If a juvenile kestrel or Peregrine's tail feather is pulled out, it grows back as an adult feather. However, a pulled wing feather may not grow back until the annual moult. Each year most raptors go through a complete moult, but large raptors such as Wedge-tailed Eagles do not and you can see juvenile feathers mixed amid adult feathers after the moult.

Summer and autumn are good times of the year for the observer to separate first-year from older raptors because first-year raptors have a clean outline whereas older birds have holes or gaps in their wings or tails where feathers are missing. Also, many just-fledged raptors have a light edge on the tail-end.

Bird killers such as Peregrines have the stiffest flight feathers; hunters of mammalian or reptilian ground prey, such as Brown Falcons, have softer, more pliable flight feathers. Owls have the softest body, flight and tail feathers because of their sound-deadening features. Peregrine wings make a loud '*swoosh*' when they dive at you, but owl wings are silent.

All raptors have water-repellent plumage to a greater or lesser degree, but this varies between species. Peregrines appear to be more waterproof than Brown Goshawks, Brown Falcons or Whistling Kites, so they can more easily hunt in the rain. Ospreys have harder feathers on their breasts and legs, the body parts that hit first when they strike the water.

Eyes and ears

Birds of prey are visual predators with eyesight directed forwards. Compared with other birds, they have large eyes. Unlike raptors, most other birds have smaller sideways-looking eyes that help the prey-bird watch for predators attacking from the side or behind.

In the raptor's eye, cones are for colour vision and rods are for black and white low-light vision and visual acuity. It is said that the 'grand champion' for visual acuity so far measured is the Eurasian Buzzard with an estimated 1 000 000 cones per square centimetre, ~2.2 times as many per unit length as in humans.

Contrary to popular belief, raptor vision is not like looking through binoculars, it is like seeing with far more detail than a human sees, like comparing a low and high-megapixel digital camera. They also see 'faster' than humans, that is, they can separate two visual images at twice the speed of humans. Muscles in a raptor's eyes are particularly well developed for focusing, increasing the curvature of the eyeball to sharpen objects close by, or flattening the eyeball to focus on objects at a distance. Humans have three types of cones for detecting colour but raptors have four and Eurasian Kestrels, *Elanus* kites and probably some other raptors can perceive ultraviolet reflections of rodent urine and faeces. They use this to locate rodent-rich meadows.

Each eye of a diurnal raptor has two *foveae*, acute focal points where receptors are especially numerous: a central fovea picks out movement, the temporal (side) fovea picks out detail that makes raptors experts at picking out defects in prey from a distance (this means that if an animal is weak or sick the raptor can pick up on this quickly and score a possible meal). As in reptiles, but unlike mammals, birds also have a *pecten* in each eye, a highly vascularised body that may feed oxygen to the eye. It is smallest in nocturnal birds, bigger in diurnal seed-eating birds, but largest of all in diurnal raptors (Figure 7.10).

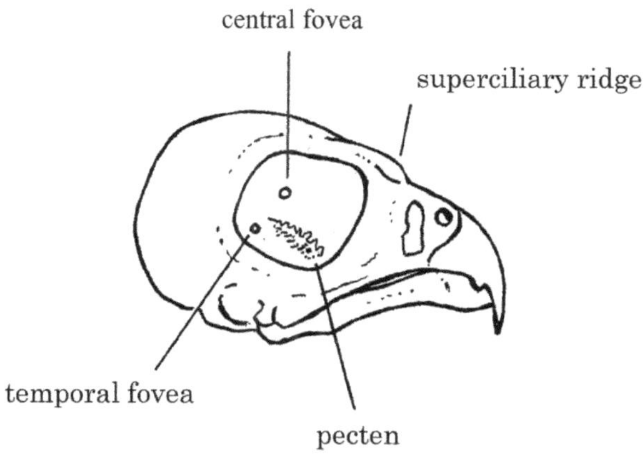

Figure 7.10 Falcon skull showing location of two fovea, pecten, and brow ridge.

Because raptor eyes are relatively large, they extend out from the face, and most raptors have a protective brow over the eye (superciliary ridge) made of bone and/or feathers (Figures 7.10, 7.11) that gives them a 'fierce' look. This structure is said to reduce excessive overhead sun glare passing through the iris and damaging the sensitive retinal surface. This bony shelf could also provide eye protection when prey flails about or the raptor pursues prey through forest, long grass or flocks of prey.

Figure 7.11 The protective brow over a Brown Goshawk's eye gives them a 'fierce' look.

Figure 7.12 Nankeen Kestrel sleeping with its face tucked into its scapulars.

They also have a nictitating membrane, a sort of semi-transparent third eyelid with brush-tipped cells that birds draw across the eye under their two other eyelids to protect it from damage (the pink fold on the inside corner of the human eye is a vestige of this membrane). Falcons, Black-shouldered Kites and other raptors have dark shading around the eye that may also reduce glare.

Diurnal raptors sleep perched upright reaching their face around and tucking it in between the scapulars on their back, not under the wing as is sometimes claimed (Figure 7.12).

Hearing is less important and less well developed than sight in diurnal raptors, though harriers have a facial disc designed for funnelling sounds, and accipiters and Brown Falcons may listen for prey in heavy cover, then pursue it.

Beaks

Raptors are designed to hunt, kill and dismantle prey. Once they have seized or disabled prey with the feet, falcons use their short, thick hooked beaks for killing prey. Unlike most other raptors, falcons have a 'tomial tooth' in the upper mandible, a spiky projection that they slip between vertebrae of their prey to snip the spinal cord. Ancient falconers considered falcons 'noble' because of this, and hawks ignoble because they lack it. Other raptors such as eagles, hawks and large kites tend not to kill with the beak; with their proportionally larger talons, they kill more by a spasmodic kneading action that stabs into the prey's vital organs and compresses the thorax, or they even start eating into the chest cavity through to the heart, though you will see raptors such as Wedge-tailed Eagles bite the back of the head of kangaroos they kill, and some eagles tear the heads off bird prey.

The beak of a raptor grows continuously so must be worn down by eating prey or dirt (raptors also swallow stones to help digestion), and you will notice that some raptors in zoos have overgrown beaks because they have no access to beak-shortening materials.

Beak size is important for dismantling large prey. Black-shouldered Kites have small beaks for dismantling smaller prey, but huge gapes for swallowing rodents whole. All raptors hunch over their prey, hook their beaks into it, and pull up with their back muscles against the talon-pinned body of the animal to rip flesh. An eagle cannot carry a whole kangaroo and a Wedge-tailed Eagle has a huge beak compared with other *Aquila* and *Hieraaetus* eagles, probably for dismantling large prey such as kangaroos and carrying away pieces.

Most hawks and eagles have oval-shaped nostrils, but falcons tend to have round-shaped nostrils. Falcons have a nipple or tubercle inside the nostril said to baffle the air at high speeds so the falcon can breathe. If you look closely at a Peregrine, Brown Falcon or kestrel these baffle systems are slightly different and their function, in truth, is a bit of a mystery, though some would say that the baffle system in the Peregrine is more complex so they can breathe while in a higher-speed stoop. In Ospreys, which dive underwater to catch fish, the nostrils can be closed, but in White-bellied Sea-Eagles, which snare fish from the surface of the water, they cannot.

The cere is the bare piece of easily cleaned skin at the base of the beak. This character is diagnostic and can be bright yellow-orange in species such as Peregrines but grey in Brown or Black Falcons, with some old male Brown Falcons eventually achieving a bright yellow cere. Raptors need to clean meat, fur, feathers and other materials from the beak after feeding. They do this by vigorously rubbing the beak on the perch or ground, first one side, then the other.

Wings

Raptors can have pointed wings for speed, rounded wings for agility, or large wings with slotted primaries for soaring. In addition, the wings contain small muscles that help alter wing shape. Looking at a bird wing, or an aeroplane wing, end-on, that is in cross-section, you can see how the leading edge bulges, the trailing edge is sharp and knife-like. Lift comes from the air flowing over the curved top of the wing from the bulging front to the trailing back further and faster than over the bottom, concave part of the wing. End-on, if the curve of wing is great, the wing has more *camber* and more lift, but the wing has more drag, so it is slower. A Whistling Kite or Wedge-tailed Eagle has such a wing, and has deeply slotted primaries to boot, so their flight action is slow, sometimes laboured. To stay aloft, they find thermals, columns of air above plains or cities, or updrafts caused by wind bouncing upwards from ridges. Though powered flight is expensive, said to be four times as expensive as gliding (Gessaman 1980), thermal flight is cheap, and can facilitate hunting and courtship flight (Peeters and Peeters 2005).

In contrast, the curve of a falcon's wing, end-on, is flatter, and most falcons have no deeply notched primaries. The falcon wing has less camber, so the wing has less lift, and the falcon has to flap more often to stay aloft; consequently the flight of a Peregrine or Hobby is fast and snappy. The flat camber and narrower wings of falcons also produce far less drag than do the broad, fingered wings of eagles and large kites, so falcons can fly and dive faster.

Wing size and shape is related to *wing-loading* – the weight (mass) of the raptor divided by the raptor's wing area. Raptors with low bodyweight compared with their large wing area, such as Black or Whistling Kites, can soar with little effort for long periods. Raptors with heavy bodies compared to small wing areas, because of narrow wings, as in Peregrines or Hobbies, or short wings as in sparrowhawks and goshawks, need to flap more to stay aloft.

These two separate non-soaring wing designs, accompanied by quicker wing beats, have evolved to allow swift flight and pursuit of prey, especially birds: 1) the short, broad wing of the woodland accipiters have deeply slotted primaries that give them tremendous manoeuvrability; 2) in contrast, the wings of most falcons are pointed with little notching and this give them tremendous speed in stoops or direct pursuit.

The *alula* or 'bastard wing' (Figure 7.13) consists of several quill feathers attached to the first digit of the leading wing, which can be extended to prevent stalling. They deploy the alula in fast glides, during stoops and when landing.

Tail

Raptors normally have 12 tail feathers and can fan these to make a broad tail or fold them tightly to make a narrow tail. The amount that a raptor fans its tail is broadly related to the raptor's wing span and the tail can provide an estimated 10% of lift when flying (Tucker 1992). The tail also acts as a rudder, most highly developed in the long tail of woodland accipiters such as Brown Goshawks or Collared Sparrowhawks that dodge through branches and barriers, but also in falcons such as Hobbies or Brown Falcons when they twist their tails to begin a stoop or pursue birds in midair. The downside of a long tail is it adds drag during flight. Among the world's large soaring raptors, the Wedge-tailed Eagle of Australia and the Lammergeyer of Africa and the Mediterranean have proportionally the longest tails; perhaps these give extra lift when soaring.

Because falcons fly faster than other raptors, they have two extra bones accessory to their tailbone called the *accessory pygostyle*. These bones help give rudder control when falcons twist and turn the tail during high-speed manoeuvres.

Figure 7.13 Alula extended on a kestrel's wings (Charles Davis).

Legs and feet

A raptor's legs need to be strong and relatively stout, for slamming into prey, pinning prey to the ground, or taking off while carrying heavy prey. Most raptors do little walking or chasing prey along the ground as their legs are too short. For example, hawks that prey on rodents and/or insects, such as the Black-shouldered Kite, kestrel or Black Kite, have relatively short legs and toes, and falcons such as Peregrines and Hobbies have short legs and longer toes for snaring birds in midair. In contrast, harriers, Brown Falcons and Wedge-tailed Eagles have longer legs because they spend more time on the ground. Also, running on such long legs helps a heavy bird like a Wedge-tailed Eagle get airborne.

Bird-catching raptors, accipiters and falcons have similar feet with long toes, but different legs. Peregrines and Black Falcons have very short legs to withstand the impact of colliding with prey at high speed; sparrowhawks and goshawks have longer legs for snatching birds from bushes or the forest canopy. Both types of bird-catchers have nipple-like pads on the undersides of their feet apparently to

Figure 7.14 Talons of the inner and hind toes on an eagle are the largest and most lethal.

grip and hold feathers. The pads on the undersides of the feet of a rodent-killer like the Black-shouldered Kite are flatter and more hemispherical.

A raptor's talons are needle-sharp, and because raptors generate an extraordinary grip, talons inflict more damage than does the beak. Talons grow continuously so they need to be worn down by use. In diurnal raptors the talons of the inner toes and hind toes are the largest and most lethal, especially in eagles (Figure 7.14), and these can stab victims to death by piercing vital organs. The smaller talons on the other toes hold quarry.

In falcons the inner and hind talons are smaller than in accipiters, eagles, harriers or kites, maybe because falcons kill prey with a bite to the back of the neck. Talons of many falcons and accipiters are like grappling hooks for snaring birds in midair; talons of eagles and large kites are more designed for pinning prey to the ground.

The flexor tendons clench a raptor's foot and hold prey (or hold a sleeping raptor on its perch). This tendon has grooves that interlock with the sheath that surrounds them ratchet fashion, so a raptor's foot can remain locked without expending energy. It also means that occasionally a raptor cannot detach from quarry and Ospreys sometimes drown attached to large fish.

In cross-section most talons are not round but hemispherical, slightly grooved on the underside so they cut like knives when they pierce quarry. For some reason, however, the talons of Black-shouldered Kites are rounded in cross-section. The talons of Ospreys are strongly curved and of about equal length, for wrapping around a fish.

Nocturnal raptors

Many animals feed at night, including kangaroos, mice and possums, but owls (Figure 7.15) have to find and capture moving prey at night that has evolved to escape from them. For this, owls need special adaptations.

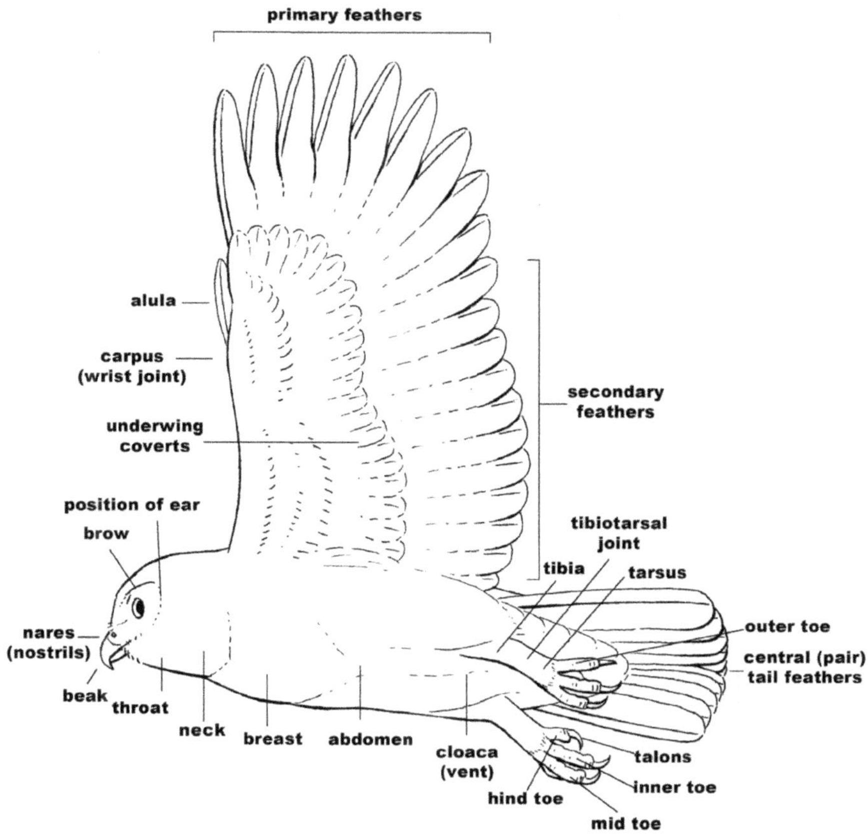

Figure 7.15 Body parts and feathers of an owl.

Figure 7.16 Owls have softer, more flexible primary feathers than most diurnal raptors.

Feathers

As mentioned above, bird killers such as Peregrines have the stiffest flight feathers of any Australian raptor and owls have the softest and most flexible (Figure 7.16).

All Australian owls are virtually silent in flight. Some owls found elsewhere, like the Northern Pygmy-Owl in North America and Eurasia, are diurnal and have hard primaries that make a considerable swishing noise in flight. Other owls have a soft, velvet-like nap on the upper surfaces of their flight feathers that absorbs sound and prevents the feathers from rubbing together and making noise (Figure 7.17).

In fact, owls have soft feathers overall, but the primaries in particular have a serrated leading edge that suppresses noise when cutting through the air. The trailing edge has softened edges that also deaden sound. Barn Owl primaries are so thin that you can read a newspaper through them (Figure 7.18).

In addition to stealth, soft wing feathers might assist hearing. The swishing sound of normal wings might deaden the acoustic system that some owls use when

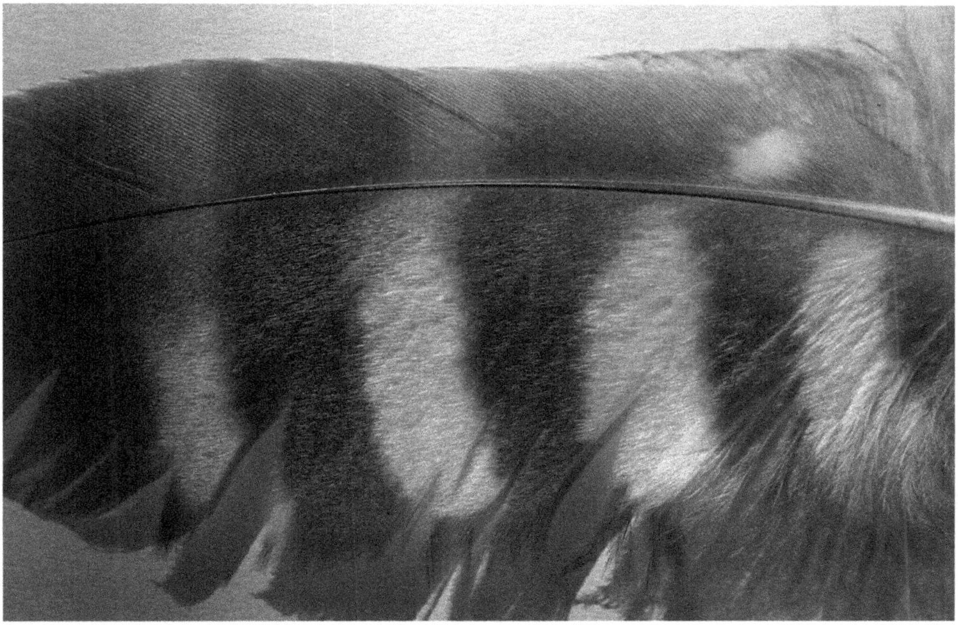

Figure 7.17 Velvet-like nap on the upper surface of a Southern Boobook feather.

attacking squeaking or rustling prey (Peeters 2007). Because of these adaptations, aircraft designers study Barn Owls to learn how to build quieter aircraft (Thomas *et. al.* 2002). These soft feathers also have soft colour tones and markings that can match tree bark and foliage, and therefore hide a roosting owl. When first spotted, an owl sometimes compresses this soft plumage and slits its eyes, as camouflage.

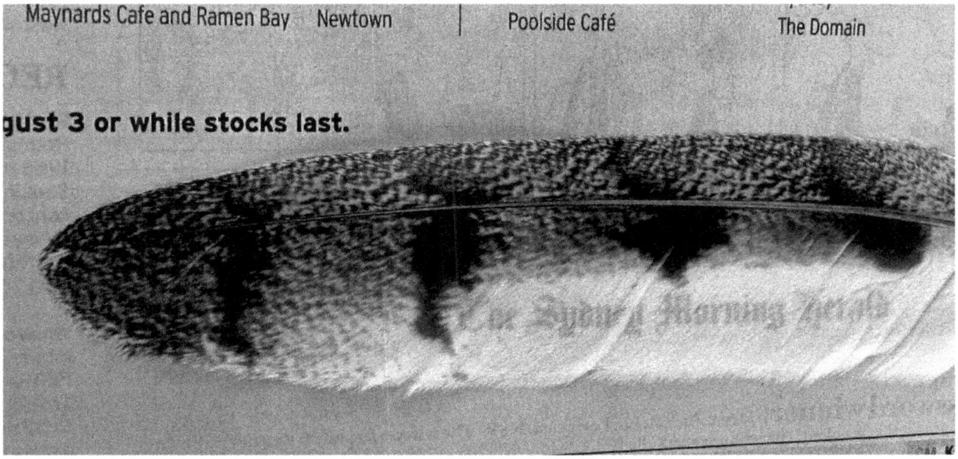

Figure 7.18 Newsprint can be read through a Barn Owl primary feather.

When cornered, they puff up their feathers to make themselves look bigger, and they bow and spread their wings and rock from side to side.

As mentioned above, feathers are dead structures. Owls spend a lot of time grooming, taking oil from a preen gland at the base of the tail and running each wing and tail feather through the beak. This gland also contains a precursor to Vitamin D that, when exposed to sunlight, is converted. Owls bathe in water or dust. Unlike most diurnal raptors, they allopreen each other as a form of pair bonding or family bonding, but also to remove ectoparasites, for example, the odd, flat-bodied Hippoboscidae blood-sucking flies that scurry like miniature crabs under the feathers of raptors. Owls reach head feathers with their claws, often scratching their heads like a dog scratches its ear. Each year most owls go through a complete moult, and looking for missing feathers or mottled appearance is a good way to separate first-year birds from older birds.

Eyes and ears

Compared with most other birds, owls have relatively large heads with very large forward-looking eyes. As in other raptors, an owl's eyes give stereoscopic perception, probably more so than in any other bird. Due to the fusing of the sclerotic ring with the skull, their eyes are simply too big to move in their sockets. They have little lateral vision (ability to see to the sides) but this probably is not such a detriment at night; lateral vision is better developed in diurnal raptors. The large size of an owl's eyes gives them excellent vision at night, similar to a domestic cat's vision and, like a cat, they can also see well during daytime. The light-gathering power of an owl's eyes has been compared to the *f* values in a camera lens (larger *f* values = smaller aperture): the widest aperture of a Eurasian Tawny Owl is *f* 1.3 while that of a human is *f* 2.1 (Marks *et al.* 1999). If you're a photographer, you know that wide apertures have less depth of field so objects in the foreground and background are out of focus, only objects in the mid-range are in focus. Smaller apertures give broader depth of field so objects in the foreground through to the background are all in focus. Owls can quickly cut pupil size and this could help to achieve greater depth of field if they are hunting, jumping to a new perch, or performing some other activity (Heinrich 1987).

Though they also have cones for colour vision, and rods for black and white low-light vision, owls have them in reverse proportions to diurnal birds: 80% rods and 20% cones, so their vision is not as sharp as that of diurnal raptors. Unlike diurnal raptors such as the kestrel (see above) owls cannot perceive UV light.

Some owls, like the Southern Boobook, have white 'eyebrows', and these are moveable; the owl has muscles to change the shape of these white markings. All owls have closable upper and lower eyelids that, with the eyebrows, add to their 'human' look. They sleep head-erect with eyes closed by these twin eyelids. As with sleeping diurnal raptors, young Boobooks can reach their face around and tuck it in between the scapulars on their back. Like diurnals, owls have a well-developed

Figure 7.19 The protective nictitating membrane covering the eye of a Powerful Owl eating a possum.

nictitating membrane, a translucent fold of skin with brush-tipped cells drawn across as a third eyelid (Figure 7.19). They activate this eyelid when struggling prey threatens injury to an owl's eyes, or when the owl is feeding prey to their own young, apparently because they fear getting poked by an outstretched flapping wing or a sharp beak. In fact, owls often kill prey with their eyes closed.

Owls' flattened faces allow their eyes to be spaced far apart. This helps them gauge enemies and prey three-dimensionally. By circling and bobbing its head, an owl sights an object from a variety of minutely different angles and receives a better picture; because of this, they can make better decisions about hunting in thick cover or catching prey in midair. A good way to discriminate young owls from adults is that young owls 'head-wind', a behaviour peculiar to owls. When owls look hard at an object, they sometimes wind their flattened faces in a plane, like a dinner plate moving in circles, first clock-wise, then back. Young owls 'head-wind' so they can better see an object, bring it into focus at their points of retinal reception (the fovea). They wiggle their whole bodies back and forth, wagging their tails when they head-wind (Figure 7.20).

Figure 7.20 Juvenile Southern Boobook head-winding (Alan Nicol).

Though owls look short-necked, they actually have 14 cervical vertebrae (humans only have seven) which allow them to easily turn and look over their shoulder, even swivel their heads 270° (Figure 7.21).

The base of an owl's skull, as in other birds, has a single condyle (knob) articulating with the vertebral column, like a ball head on a tripod. It is the owl's flattened face and widely spaced eyes that make it appear as though its head is mounted backwards when it turns its head 180° (Peeters 2007). Young owls like to turn their heads upside down to gain a different perspective. Other raptors, especially young falcons, also turn their heads upside down to watch butterflies or passing birds, and they can look backwards over their shoulders, but they don't 'head-wind' as owls do.

Owls have excellent hearing. If you pucker your lips and do a mouse-squeak at a Boobook at night, it will almost always turn and look down at you. Again, their widened faces space their ears from one another and so allow them to triangulate sounds. Since the 1960s we have known that Barn Owls can capture prey in total darkness. *Tyto* owls have asymmetrically placed external ear flaps, one higher on the side of the head than the other, and a facial disc that funnels sounds into these ears, so Barn Owls can triangulate the sound of scurrying or squeaking rodents and nail them with pinpoint accuracy. In Barn Owls the left opening on the side of the head is above eye level but points downwards; the right ear is below eye level

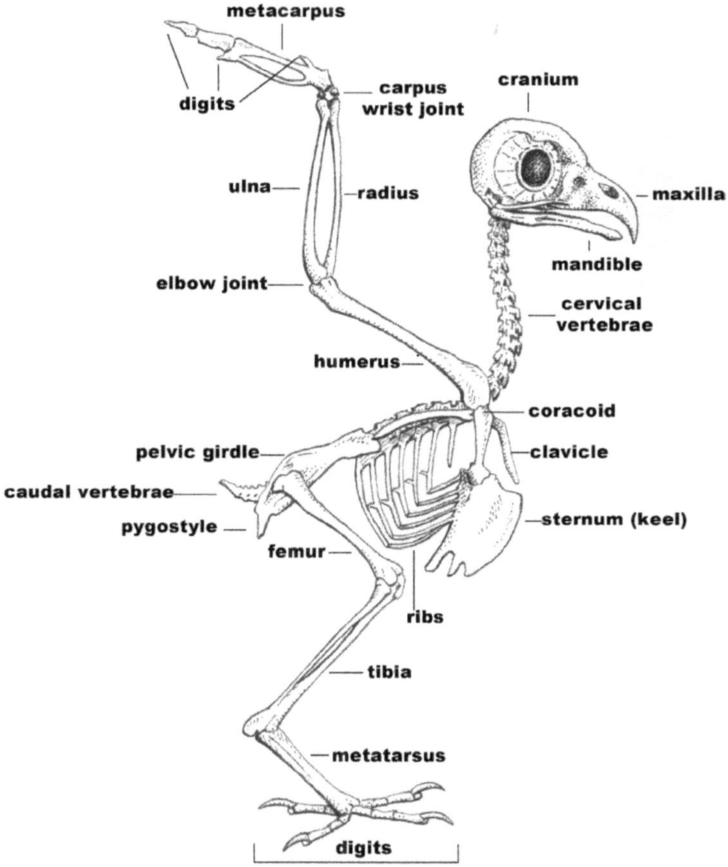

Figure 7.21 Typical skeleton of an owl.

pointing upwards. Konishi (1973, 1983, 1993) experimented with captive Barn Owls by blocking one ear, or removing feathers from the facial disc to make the disc incomplete. In the dark, the owls missed their target. Contrary to some claims, Barn Owls probably have relatively small eyes (compared to other owls) because they hunt more open areas, not solely because of asymmetrical hearing. Sooty Owls, also with asymmetrical hearing, have the largest eyes of any Australian owl.

Ninox owls, such as Powerful Owls, don't have asymmetrical ears and tend not to hunt ground prey in dark conditions, relying instead on sharp eyes to spot and pursue arboreal mammals and roosting birds in the forest canopy, and chase flying insects such as beetles in midair. Boobooks are an exception in that they take flying bats and insects and roosting birds, but they also hunt ground prey. Boobooks do not have asymmetrical ears but they have a facial disc for funnelling sound and they use hearing to hunt prey, including ground prey.

Hearing is also crucial because owls, in contrast to most diurnal raptors, are vocal and they need to recognise the nuances of various calls and callers, and the identity of conspecifics. As in songbirds, owls proclaim their territories, attract mates, and challenge neighbours through vocalising. Territorial calls, like the *boo-book* calls of the Southern Boobook, or deeper *who-who* of the Powerful Owl, are used against conspecifics, and as contact calls to mates.

Beaks

Owls have rather short, decurved, powerful beaks, partially hidden in rictal bristles (Figure 7.22) that are almost like a bristly moustache. They have no tomial tooth, as falcons do, but they break a victim's neck in a similar manner. Powerful Owls have particularly long, large and powerful beaks for an owl, probably for killing and dismantling large prey.

Wings

Like most birds, owls have a lightweight skeleton for flight, with bones supported internally by struts (Figure 7.21). Their wing bones are relatively long, associated with a broad surface area, which gives them a low wing-loading compared to birds like falcons or swallows (think of a biplane compared to a jet). The wings contain small muscles for altering the wing shape.

Figure 7.22 Rictal bristles around the beak of Southern Boobook.

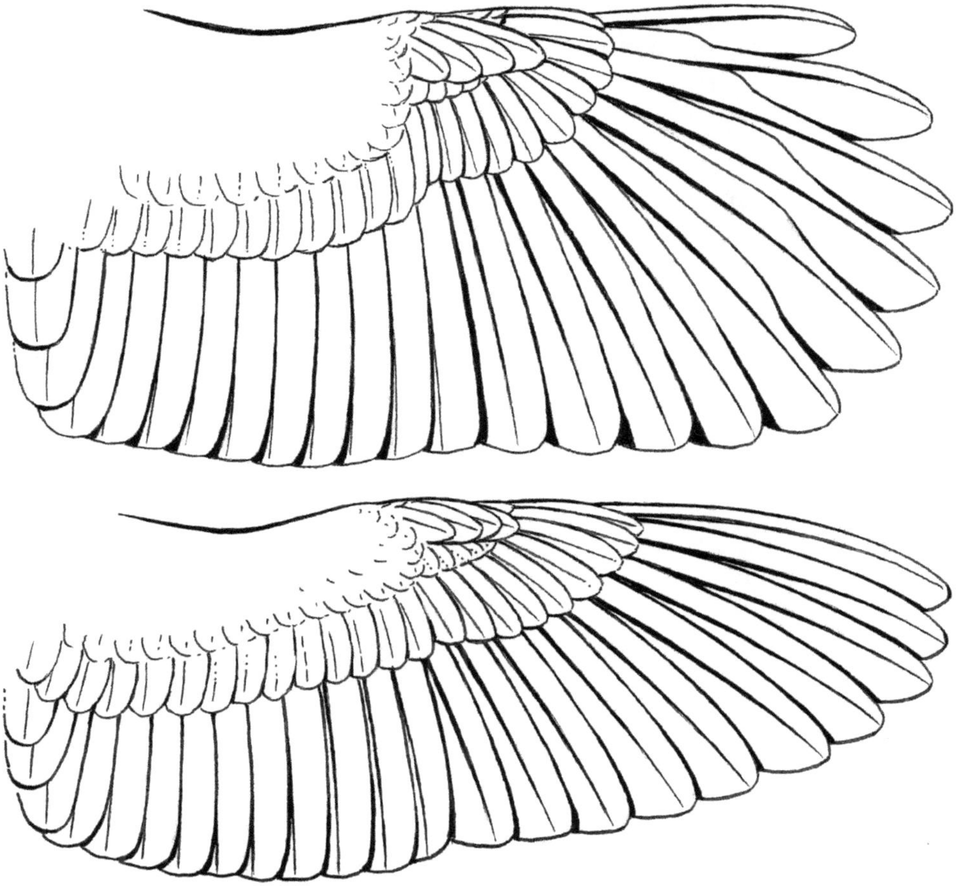

Figure 7.23 Wing shape of a forest owl (Powerful Owl, top) and open-country owl (Barn Owl, bottom).

Most owls have a relatively small sternum compared to diurnal raptors such as falcons, because owls lack the chest musculature of diurnal raptors and do not reach the speeds that diurnal raptors do. But owls can easily lift and carry heavy prey, even in still air, something that Hobbies or Peregrines with higher wing-loading and narrower wings can't easily do without a strong headwind; owls don't need a strong headwind to carry heavy prey. Broadly speaking, owls with comparatively short, broad wings, such as Powerful Owls and Southern Boobooks, are still-hunters that often choose a perch in woodland and wait for prey. Owls with longer wings, such as Barn Owls, have lower wing-loading that helps them actively quarter over open ground, searching for prey (Figure 7.23).

Tail

Most owls have short tails, especially compared with accipiters and some falcons, and *Tyto* are no exception. But large *Ninox*, such as Powerful Owls, have long tails,

Figure 7.24 Unlike the talons of most diurnal raptors and some owls, the talons of a Southern Boobook are all about the same length.

clearly evolved as a rudder for forest hunting. *Ninox* seem to be nocturnal versions of the diurnal accipiters.

Legs and feet

The foot bones and tarso-metatarsus of owls are relatively short and stout, with highly developed curved talons on all four toes for efficient killing and carrying of large prey (Figures 7.24, 7.25). Diurnal raptors have prominent hind and inner talons for stabbing prey to death, but in owls the hind talon is not so prominent, maybe because they kill prey with a bite to the back of the neck.

Barn Owls have a comb-like claw on the middle toe (pectin) said to be used for grooming (Figure 7.26) but it probably has a different function.

Most owls differ from most diurnal birds of prey, like falcons, in that their legs are feathered down to the toes. Some researchers say this is for protection against prey that bites back, such as rats, but small owls that catch insects also have feathered legs, and most diurnal raptors such as goshawks, sea-eagles or falcons do not need feathered legs even though they take and subdue large, dangerous prey.

Reversible outer toes in owls

The biggest difference between the feet of owls and those of diurnal raptors, except Ospreys, is that owls have a 'reversible' outer toe. They can perch with three toes

Figure 7.25 Tawny Frogmouths are not raptors and have relatively weak feet, so they kill with their beaks.

Figure 7.26 Comb-like claw on the middle toe of *Tyto* owls.

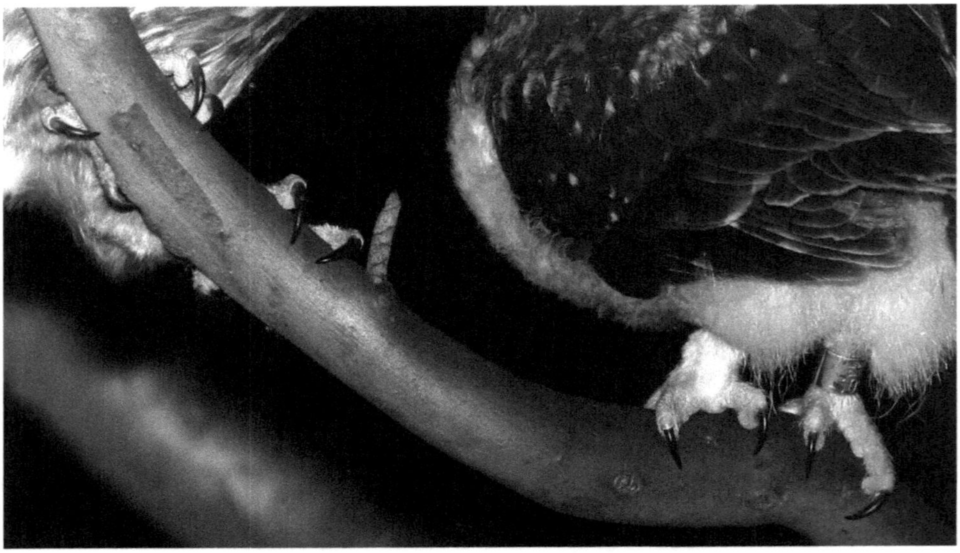

Figure 7.27 Zygodactyl toes arrangement in perched Southern Boobooks.

pointing forward and one pointing back, the usual 'anisodactyl' manner of most birds, including falcons, eagles and kites, or they can swing the outer toe behind them, parallel to the rear toe, and perch with two toes pointing forward and two facing backwards, in a 'zygodactyl' arrangement (see Figures 7.27 and 7.28).

Some have argued that owls have evolved this arrangement to better clutch and carry heavy prey. It is possible, however, that owls are taxonomically close to birds like woodpeckers that already had two toes forward and two toes backward and the hunting advantage for owls evolved second, though Olsen (2011a) questioned whether this advantage is real. All diurnal raptors except Ospreys seem to get along fine without it. Owls could be described as having a zygodactyl arrangement from which a reversible outer toe swings *forward* into an anisodactyl position for holding small prey. If you watch Southern Boobooks or Little Owls eating invertebrate prey in one foot, they use an anisodactyl toes arrangement.

Some have argued that the Australian Black-shouldered Kite can rotate the outer toe to a zygodactyl arrangement *contra* Johnsgard (1990) and Dunk (1995), citing photographs in Cupper and Cupper (1981), Hollands (1984) and comments in Negro *et al.* (2006) as evidence. However, photos of Black-shouldered Kites I looked at, in collaboration with Geoffrey Dabb and Susan Trost, did not show zygodactyl arrangement.

Owls commonly perch using a zygodactyl toe arrangement (Figures 7.27, 7.29, 7.30). But hunting owls seldom catch ground prey using a zygodactyl arrangement, they mostly catch prey with toe 1 opposing toes 2 and 3, with the outer toe, 4, at a 90° angle to toes 1 and 3, an arrangement that could be term 'part-zygodactyl', or 'lateral 4' (Figure 7.28, middle row). This arrangement is

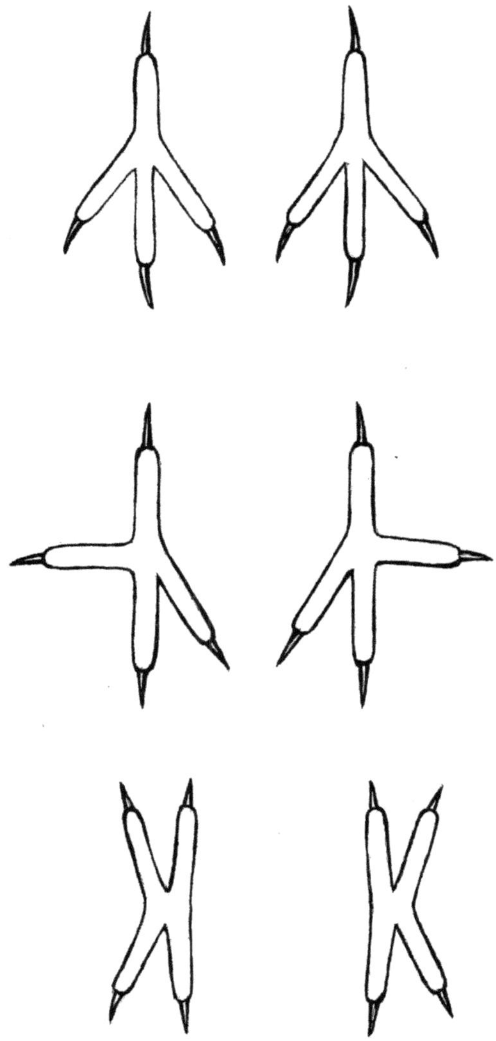

Figure 7.28 Toe arrangements in raptors: *anisodactyl* top, *part-zygodactyl* (*'Lateral 4'*) middle, *zygodactyl* bottom.

superior to a zygodactyl arrangement because it gives a wider lateral spread of talons, but the inner toes (# 2) don't spread and overlap to get in each other's way. It is interesting that *Elanus* kites also use this form of toe arrangement, part-zygodactyl instead of zygodactyl, probably because it is superior to the anisodactyl or zygodactyl arrangement. Geoffrey Dabb has pointed out that harriers may also use it. Part-zygodactyl may have developed in species that already had either facultative (optional) zygodactyl toe arrangement (owls, Ospreys) or anisodactyl toe arrangement (*Elanus* kites and harriers). That is, it may have arisen independently across taxonomically unrelated raptors.

Figure 7.29 Owls commonly perch using a zygodactyl toe arrangement, but eat small prey using an anisodactyl toe arrangement.

Figure 7.30 Southern Boobook midway between moving its toes from anisodactyl to zygodactyl toe arrangement.

8

Food and hunting

More differences between overseas and Australian raptors

In 2011 Sue Trost and I were on the south rim of the Grand Canyon with Chris Parish, from the Condor Rescue team in Arizona. We watched a Turkey Vulture and tried to locate the radio-tagged California Condor Chris had detected.

I related a story from 1969. On the 20th of July in wilderness on the shore of Lake LeBarge in the Yukon Territory, the lake where the poet Robert Service had cremated Sam McGee, I noticed two immature Bald Eagles eating something on the beach. I left my friend John at the cabin and walked half a mile down the shore and flushed the eagles. Surprisingly, there was nothing on the spot where the eagles had been feeding, or so it seemed, and I looked around more closely. Suddenly I saw a hand waving in the water, attached to an arm and a body, and realised I had been standing over a dead man lying face down mostly submerged in the water. He had only a small tassel of hair left on the crown of his head and he smelt sulphurous, the rotten eggs smell of ammonia. I guess I hadn't seen him, even though his right leg was at my feet, because you see what you expect to see. The eagles couldn't get through the boot covering his foot that stuck out of the water onto the beach so they had torn open his pants leg and these USA national emblems, close relatives of our White-bellied Sea-Eagles, had been eating the dead man's calf.

The two Condor watchers listened, but they didn't seem too impressed with the story. They said that about three people a year die in the Grand Canyon from falls,

Figure 8.1 Unlike many countries, Australia has no vultures, such as this Turkey Vulture.

suicides, being lost without food and water, whatever, and Turkey Vultures (Figure 8.1) or Condors find them and eat them. A few weeks later, Pete Jenny, Director of the Peregrine Fund in Boise, Idaho, told us that people searching for the lost adventurer billionaire Steve Fossett (he went down in an aeroplane somewhere in California or Nevada, and it took over a year for his remains to be found) had approached the Peregrine Fund and asked if they could put radios on Condors to track them in case they discovered Fossett's body and started to eat him.

I can't think of any such stories for Australian raptors. Unlike Australian raptors, many vultures locate prey by smell. In Washington State you see Turkey Vultures on road-killed deer. Breeding Golden Eagles rarely feed on carrion, but their close cousin, the Wedge-tailed Eagles, and breeding Whistling Kites, Black Kites and ravens take the place of vultures in Australia. In India human bodies are put up on frames to be consumed by White-backed Vultures but this system came apart when the veterinary drugs diclofenac and ketoprofen (in treated cattle) poisoned these vultures and most of them died after eating dead cattle. North American and European vultures migrate to Central and South America and Africa respectively, but Australian raptors have no giant grasslands to migrate to, and vultures did not seem to make it to Wallace's Line. If you look at Geometric Mean Prey Weight (a calculation of the average weight of prey species taken by a raptor) compared to the average weight of a raptor to obtain a prey/predator index

Table 8.1 Comparison of raptor weight, geometric prey weight (GMPW) and prey/predator ratio of some Northern Hemisphere and Australian raptors

Species	Region	Raptor weight (g)	GMPW (g)	Prey/predator ratio	Reference
Red-tailed Hawk	Idaho	1126	135	0.12	Steenhof and Kochert 1988
Whistling Kite	Canberra region	870	227	0.26	Fuentes *et al.* 2005
Little Eagle	Canberra region	868	249	0.29	Olsen *et al.* 2010a
White Bellied Sea-Eagle	Canberra region	2865	887	0.31	Olsen *et al.* 2006a
Bald Eagle	North America	4683	657	0.14	Marti *et al.* 1993
White-tailed Sea-Eagle	Europe	703.2	4793	0.15	Marti *et al.* 1993
Wedge-tailed Eagle	Canberra region	3466	1291	0.37	Olsen *et al.* 2010a
Golden Eagle	Idaho	4308	690	0.16	Steenhof and Kochert 1988
Brown Falcon	Victoria	584	77	0.13	McDonald *et al.* 2003
Prairie Falcon	Idaho	648	97	0.15	Steenhof and Kochert 1988

(Table 8.1), Australian eagles and large kites seem to take larger prey in proportion to their body size than their congeners or similar-sized raptors overseas. This could be partially because many breeding Australian raptors double-task, take carrion as well as live prey.

In Australia we're so accustomed to a landscape without vultures that we don't see it as odd. It surprises researchers in the US and Europe when we tell them that Australia has no raptor species that feed solely on carrion. We don't have stories of White-bellied Sea-Eagles eating human carrion, though they do take seals and other mammalian carcasses. Raptors such as Wedge-tailed Eagles (Figure 8.2), White-bellied Sea-Eagles and Whistling Kites tend to use some carrion during breeding, giving them a higher prey/predator ratio than non-vulture raptors overseas that don't do this. So, in regard to prey and foraging, there are similarities to Northern Hemisphere raptors, but also differences; when drawing theories and research protocols from Northern Hemisphere raptors this needs to be borne in mind.

Another example: breeding accipiters in the Northern Hemisphere such as Cooper's Hawks, Sharp-shinned Hawks and Eurasian Sparrowhawks take few or no insects compared to high country Brown Goshawks and Collared Sparrowhawks (Figures 8.3 to 8.5), both of which take many insects (see species accounts in Appendix 1). No owl weighing more than 300 g studied around the world regularly takes insects in midair as Australian *Ninox* owls do.

Figure 8.2 Unlike in the Americas, Africa, Asia and Europe, all carrion-eating raptors in Australia, such as Wedge-tailed Eagles, regularly take live prey (Felicity Hatton).

Merlins in Montana and Canada hunt small birds over grasslands and prefer low stick nests, often the roofed nests of Black-billed Magpies, in hawthorn *Crataegus* bushes. Some Merlins nest on the ground. The slightly larger Australian Hobby also hunts small birds but often over woodland and forest canopy, and they catch many more insects than Merlins, choose the highest available nest in a eucalypt and never nest in low bushes or on the ground. Many of the invertebrate prey taken by Australian raptors are linked to eucalypts so, as in the discussion of cliff nesting in Chapter 2, these examples show how eucalypts are crucial to breeding Australian raptors. Also mentioned above, eagles, some falcons and large kites in the Northern Hemisphere high country have terrestrial gallinaceous birds such as the grouse and ptarmigan, partridges *Perdix* and *Alectoris*, marmots *Marmota*, ground squirrel *Spermophilus* (Figure 8.6) and hares *Lepus* as large prey species. There are no counterparts in the Australian high country, although there are introduced hares. The study of food and hunting in Australian raptors needs to be studied in this context – raptors are predominantly meat-eaters, but the source of meat in Australia can be different from that found elsewhere.

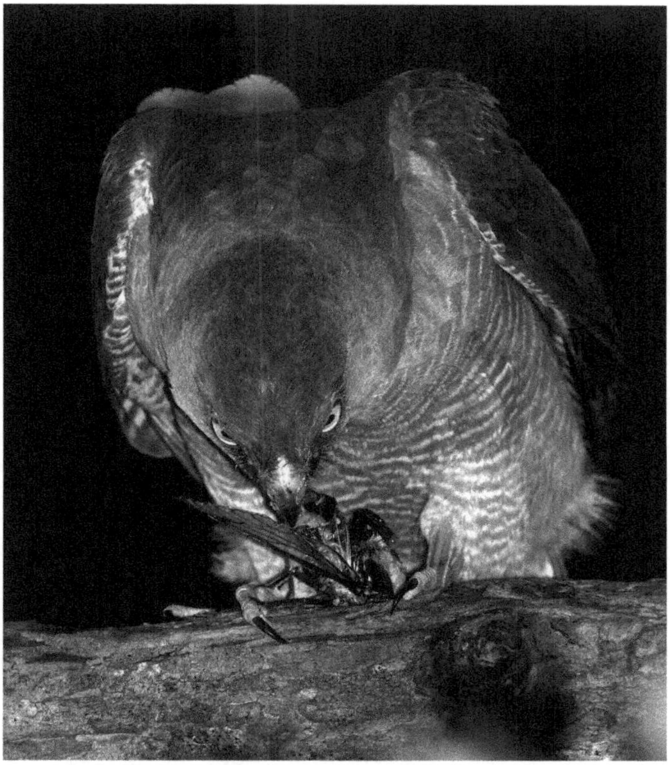

Figure 8.3 Unlike North American accipiters, Collared Sparrowhawks eat many insects, such as this Black Prince Cicada.

Figure 8.4 In Australia, Hobbies, accipiters and *Ninox* owls commonly catch invertebrates like cicadas.

Figure 8.5 Unlike most owls overseas, Australian *Ninox* regularly catch insects in midair with their feet and transfer these to their beaks.

Figure 8.6 Unlike Australia, other countries have abundant diurnal, winter-hibernating rodents, such as this Townsend's Ground Squirrel in Idaho, that serve as major raptor prey.

Meat, compared with leaves, grass and other vegetable foods, is easier to digest, mostly free of toxic or distasteful compounds, lacks the hard cell walls that plants have, and contains more protein and fat per gram than plants do. The problem in most environments is that meat is scarcer than plants and, unless it is dead, meat runs away, hides, tactically uses cover and can be hard to catch, especially if the source of the meat can fly. Prey can use cover to draw the pursuing raptor into a costly accident. Raptors mostly capture prey, including insects, with their hind limbs, not their beaks, so they need to swing the legs forward while on the chase, a difficult manoeuvre where some raptors have to tilt the flying body upright just as they strike prey (Goslow 1971). This could make them more at risk of collision. As expected, the number of swift-flying falcons or woodland accipiters ending up in rehabilitation centres is out of proportion to their numbers in the wild. Figures 8.7 and 8.8 show common collision injuries to raptors.

In spite of these risks, most Australian raptors seek living, moving prey instead of carrion. Unlike plant food, meat spoils quickly, so it cannot be stored for long periods like seeds or grasses, though it can be cached for short periods in cool weather.

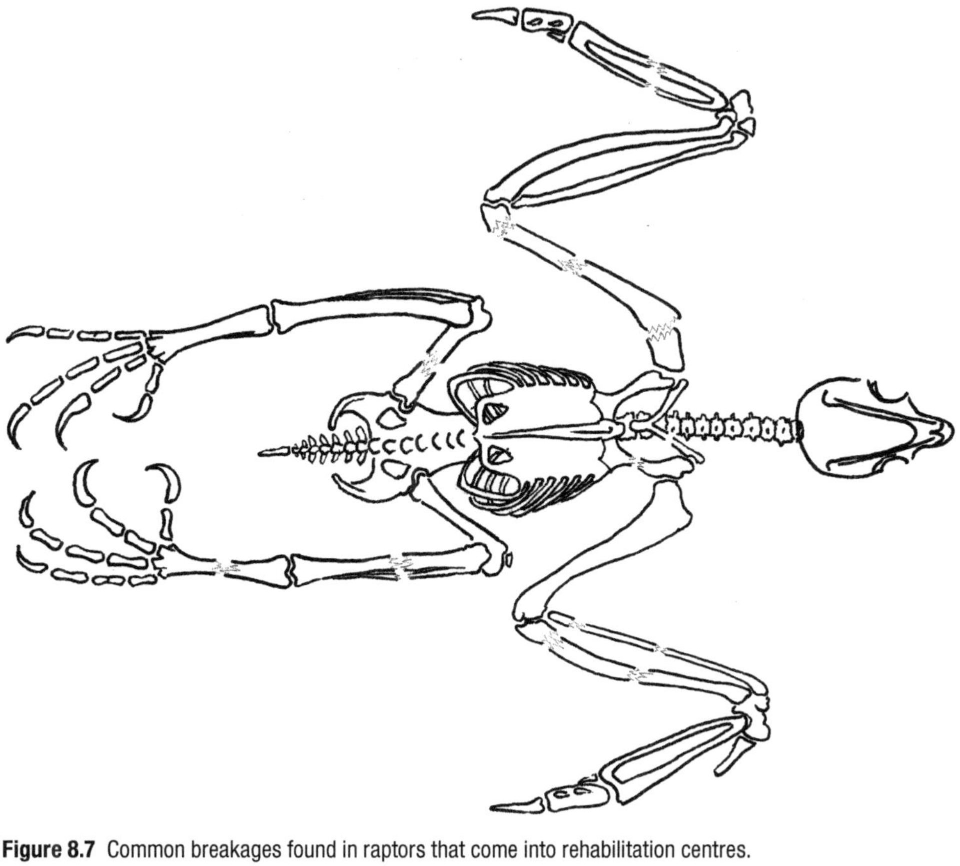

Figure 8.7 Common breakages found in raptors that come into rehabilitation centres.

Figure 8.8 This Brown Goshawk caught in a fence was rehabilitated and released (Kym Bradley).

The high value of meat is reflected in the behaviour that raptors display after they catch one of these large packages of food. Most raptors eat on the ground and they 'mantle' by spreading their wings and tail to hide their prey, particularly from raptors of the same species. Sometimes a raptor will brace its open wings against the ground when hooking its talons and beak into the prey and pulling to dismantle it, which looks something like mantling but has a different function. Some raptors, for example goshawks, tend to drag the prey under a bush, whereas Hobbies snatch and grab prey, remain airborne and, unlike Peregrines, do not come to the ground with it but eat it on a limb. Many raptors catch and eat invertebrates in midair. All of these behaviours protect the valuable meat packages from robbery, but not always successfully; robbery is fairly common. Raptors often 'prepare' the prey before eating it or carrying it to the nest. Bird-eaters commonly pluck bird prey and Wedge-tailed Eagles dismantle or partially skin large mammalian prey such as kangaroos. Excess prey is cached under a bush, in a tree or in another nest, but large *Ninox* owls such as Powerful Owls are unique because they perch on uneaten prey for the day.

Bird meat is a more nutritious food than some mammal meat, like rabbits, so birds are energetically efficient prey for raptors to eat, but harder to catch. Peregrines, Hobbies and sparrowhawks are essentially bird-catchers. Peregrine Falcons are designed for high-speed hunting and very few birds can outfly them. How raptors hunt and subdue prey, more than any other characteristic, draws humans to falcons, eagles, hawks and owls – to study them, use them as symbols of nations, or in falconry to observe and control this rapid pursuit of quarry. Peregrines fold their wings and hurtle earthward at breakneck speed. Wild falcons certainly do stoop at birds flushed by humans, machines, birds or mammals but this probably isn't the falcon's most common way of hunting. Kenward (1990) points out that raptors with a choice of hunting techniques use the method that costs the least energy.

Below, we examine some of the techniques and strategies used by hunting high country raptors. Hunting is difficult to study so there are few quantified data for Australian raptors. To understand their feeding ecology, it is important to understand which techniques a raptor can and cannot use.

Search and attack techniques

Search techniques

Nick Fox (1977) defined hunting as 'when a bird of prey searches for, locates and attacks prey'. He outlined seven **search techniques** that raptors use to locate prey:

1. Still-hunting
2. Fast contour-hugging flight
3. Soaring and prospecting
4. Slow quartering
5. Stalking
6. Listening
7. Flushing from cover.

These are summarised below and applied to high country raptors.

1. Still-hunting

A raptor perches for a period of time at a suitable vantage point, like a tree, power pole or cliff, and audits the surrounding area for unpredictable prey. It may switch short-stay perches during a hunt. This is the most common search technique used by high country raptors. Nick Fox outlined three main criteria for a suitable *still hunting* perch:

(i) good all-round view of an area well populated with prey species;
(ii) concealment from prey by foliage or broken background; for example, prey would be unlikely to see a raptor perched between the prey and the sun, or on a cliff face;
(iii) elevation so the raptor can slip off the perch into a *glide attack* without warning the prey with wing beats.

Diurnal and nocturnal raptors in the high country often launch their attacks from trees. Raptors such as owls, Grey and Brown Goshawks and Collared Sparrowhawks ambush prey unseen. Peregrines especially use cliff tops, kestrels use trees, fence posts and buildings, and many raptors use powerlines, poles, or any elevated position. Peregrines or kestrels perching on cliffs are more difficult to see than raptors perched in trees.

2. Fast contour-hugging flight

The raptor flies rapidly a few metres above the ground or above the tree canopy following the contours of the terrain, covering large areas and searching for prey surprised by the sudden appearance of the raptor. *Fast contour-hugging flight* is a highly active form of searching that uses considerable energy compared to *still-hunting*. Peregrines, Hobbies, goshawks, sparrowhawks, and eagles use *fast contour-hugging flight* to cover huge areas, flying at speed in straight lines, zigzagging or in broad arcs.

3. Soaring and prospecting

Raptors remain airborne at height, circling and searching the area beneath them. This form of searching is particularly well developed in raptors with broad, soaring wings that hunt ground prey like mammals and reptiles. Eagles use *soaring and prospecting* when searching for larger prey, accipiters when searching for birds or mammals, large kites when searching for carrion and other prey. Peregrines use *soaring and prospecting*, but bird prey is likely to freeze if it sees a falcon above them. Mediating flushing agents, such as humans or machinery, frighten starlings or other birds into the air so that Peregrines can take them. Raptors use *still-hunting*, or *soaring and prospecting* (Figure 8.9) for prey that is flying over the forest canopy or over open roads, clearings or dams and catch it before it can dive under the forest canopy to water. Accipiters use *soaring and prospecting* to pursue prey into and under the forest canopy.

Often a *soaring* raptor is not hunting; it is sending a message, using territorial display against conspecifics, making itself conspicuous to enemies or potential mates and, therefore, making itself conspicuous to prey. Some observers incorrectly interpret all soaring as hunting.

Figure 8.9 Whistling Kite prospecting (Roger Williams).

Kiting is similar to soaring. The raptor hangs on the wind with little or no wing movement, except slight winnowing, and little or no forward movement. Often the alula is projected forward for extra lift. This provides a solid platform for the raptor to scrutinise the ground below. The whole body of the raptor acts as an aerofoil, the camber changed by raising and lowering the tail. *Kiting* is more energy-efficient than *hovering* (see below).

4. Slow quartering

Raptors fly slowly, often low, over grassland, searching and sometimes listening for prey. The ground is thoroughly examined for rodents, birds or insects by quartering back and forth over one area. Kestrels, Black-shouldered Kites and Brown Falcons may search an area, then swing into a *hovering* position (Figure 8.10), carefully scrutinising the ground below them using hovering to create a stationary platform, like searching the ground from a stationary helicopter versus fixed-wing aircraft.

This technique requires buoyant flight and a light wing-loading as found in Swamp Harriers, Barn Owls, Black-shouldered Kites and other rodent-eaters. Raptors such as Peregrines, with their heavier wing-loading, can seldom use *slow quartering* efficiently. Accipiters hunt in woodland and seldom use quartering or hunt over open grassland.

5. Stalking

Only a few raptors hunt by *stalking* prey on foot. Brown Falcons hunt locusts on foot and commonly search for yabbies and other invertebrates such as Little

Figure 8.10 Hovering Black-shouldered Kite (Geoff Dabb).

Basket Shells (*Corbula stolata*) (Figure 8.11), and Whistling Kites hunt on foot along water edges.

Brown Falcons also hunt snakes by stalking and attacking as the snake strikes at the falcon's outstretched wings; they grasp the snake with their heavily armoured feet. Tame Brown Falcons can be taught to catch rubber snakes at bird shows (Figure 8.12).

Unlike Powerful Owls, Southern Boobooks commonly hunt invertebrates such as crickets and grasshoppers on the ground at night, and accipiters, with their long legs, sometimes do this by day.

Raptors such as Peregrines, with their short legs, do not use this technique. They seldom run after prey and catch it on the ground, and Hobbies tend not to come to the ground at all.

6. Listening

Owls, harriers and accipiters commonly listen to locate prey. As with some owls, harriers have facial discs used to funnel sound. Falcons are rarely recorded using this technique. There is no evidence that eagles, Peregrines or Hobbies use this technique but this requires further research.

Figure 8.11 Breeding Brown Falcons are generalist predators that fill a similar niche to Northern Hemisphere *Buteos*. They even take molluscs such as Little Basket Shells.

7. Flushing from cover

Accipiters, Southern Boobooks and Brown Falcons commonly enter bushes or the tree canopy searching for prey, but other falcons seldom do this. Peregrines seldom enter cover and have difficulty hunting inside thick vegetation.

Attack techniques

Seizing prey from above or below, outside of the prey's line of sight, gives raptors an advantage. When a hawk or eagle is attacking prey, its feet are often held under its tail, and when it hits the prey the raptor flips the upper body backwards, the lower

Figure 8.12 This Brown Falcon at the Western Australia Bird of Prey Centre instinctively runs down and attacks a rubber snake (Yvonne Sitko).

body and tail forward, and flings the feet forward to increase the velocity and impact. In contrast, large falcons in a stoop typically hold the feet more forward, angled under the breast, then shoot their feet out and the prey is struck with a massive blow from the falcon's open feet. The blow kills or stuns the prey, tearing off feathers that float away in a plume. Fox (1977) listed four attack techniques; I have added a fifth:

1 Direct flying attack
2 Tail chasing
3 Glide attack
4 Stooping
5 Feet-first drop.

1. Direct flying attack

The raptor flies directly at sighted prey. This is the technique used most often by high country raptors. Peregrines and Hobbies often make repeated *direct flying attacks* at passing flocks of parrots or pigeons. If they fail to capture quarry they often return to the perch and wait for another passing flock. *Ninox* owls such as Boobooks and Powerful Owls make *direct flying attacks* at passing bats and moths. Accipiters attack prey in midair or in cover, eagles often make a long angled dive at prey on the ground.

2. Tail chasing

The raptor chases the quarry over a long distance, attempting to overhaul it, and waits for the prey to make a mistake, or wears it down. Peregrines use this technique to catch Galahs, Feral Pigeons, Common Starlings or other flock birds that fly in the open and can escape high into the sky instead of diving for cover; hobbies use it for smaller prey such as birds and bats. Accipiters and eagles tail chase for short distances but lack the overall speed and endurance of falcons. Australian owls seldom use this technique, though diurnal Northern Hemisphere owls such as Snowy Owls *Nyctea scandiaca* do.

3. Glide attack

This attack is used against unsuspecting but potentially mobile prey, usually launched from a perch, and designed to get the raptor to the prey without forewarning it. The raptor slips off the perch with wings partly folded and glides, without wing beats, or with slight flickering, in a straight line towards the prey. As the raptor's speed increases, it may fold its wings tighter to its body and then open its wings just as it strikes the prey. This technique, used by most high country raptors, including *Ninox* owls, makes the raptor almost invisible to its prey.

Fox differentiated the *glide attack* from the *stoop* (see below) in that a *glide attack* is low and not repeated; the *stoop* may be repeated. Peregrines and hobbies use *glide attack* on flying birds and may follow it up with repeated short *stoops*; Brown Falcons, eagles and goshawks use it against rabbits and other prey.

4. Stooping

Raptors dive at prey from a soar or from a perch and attempt to strike or grasp it in midair. Peregrines are best known for this technique. *Stooping* is also used by cooperative hunting pairs of falcons, often with the female *stooping* at individual birds that the male has split from a flock, or vice versa.

When a Peregrine strikes a bird from a *stoop,* its toes are extended to deliver a smashing blow that stuns the prey and makes it easier to dispatch. In contrast, hobbies snatch prey with all four toes and talons, and carry it to a safe place where it can be killed and consumed.

5. Feet first drop (not in Fox 1977)

Raptors spread their wings and tail and slowly parachute onto ground prey. Kestrels, Black-shouldered Kites and Brown Falcons, all of which have low wing-loadings, use this technique from a hovering position. Falcons such as Peregrines or Hobbies with heavier wing-loadings cannot use it, but *Tyto* (Figure 8.13) and *Ninox* owls can. Prey usually can't evade quickly enough as they are in long grass or cover.

Figure 8.13 Barn Owl showing feet-first drop and 'lateral 4' toes arrangement when grasping prey.

Assessing raptor food

Many papers give prey species and the percentage of each prey species comprising a raptor's diet. How do we know this? How do we know that Peregrines take starlings but few rabbits or carrion, and Wedge-tailed Eagles take rabbits but few starlings? Biologists follow certain protocols. For example, we know that Powerful Owls prefer arboreal mammals such as Greater Gliders, Sugar Gliders, and Ringtail and Brushtail Possums because we find pieces of these animals, legs, tails and skulls (*orts*), in Powerful Owl nests or under day roosts. Because owls, eagles and some kites sometimes consume the heads of their prey away from the nest, or gulp small animals whole, biologists analyse regurgitated pellets (also called *castings*), and these can be collected from the nest, under the nest or under the day roost, for example where a male eagle perches and dismembers prey before bringing it to the nest. Unlike diurnal raptors, owls have no crop for storing food, so they may need to eat more often and store prey or perch on it for repeated meals. Owls also differ from most diurnal raptors, such as goshawks (but not Black-shouldered Kites), which have stomach acids 20 times as powerful as the acids in owls' stomachs and can metabolise some bones they swallow. This benefits owl biologists because the castings contain intact large and small bones, whole skulls, feet and fur that an owl has swallowed. It disadvantages goshawk biologists because the casting might have few bones or feet; the casting can look like a tight ball of grey felt containing few or no bones, and it is difficult to identify the feather particles when you break apart the casting. You see a lot of *unidentified bird* in lists of accipiter prey. A Brown Goshawk casting or a Boobook casting in similar habitat are about the same size, but they look different. The owl casting often contains chunky bits of bone and the chitinous exoskeletons of insects. Black-shouldered Kite castings look similar, but contain mostly rats and mice. With clunky owl castings biologists can more easily identify prey species and count prey items. Goshawk biologists might need to rely more on pieces of animal (orts) dropped under a nest or roost. In containing visible whole bones, bits of fur, teeth, skulls and beaks, falcon and eagle castings (Figure 8.14) fall somewhere between owl or Black-shouldered Kite castings and accipiter castings.

Biologists disagree about whether orts (Figures 8.15 and 8.16) or castings best represent what an eagle, kite, falcon, hawk or owl is eating because sometimes a prey species might show up in castings but not in orts, or the other way round. Recommendations in the scientific literature for collecting unbiased information on eagle diets are contradictory. The Catalonian biologist Joan Real (1996) concluded that pellets were more representative than orts for analysing the diet of Bonelli's Eagles, but Andy Sharp *et al.* (2002a) ignored pellets of Wedge-tailed Eagles in their analysis and relied solely on orts. Australian researchers might cite Real if they want to analyse pellets and ignore orts, or cite Sharp *et al.* if they want

Figure 8.14 Casting from under a Wedge-tailed Eagle nest.

Figure 8.15 Prey remains (orts) of kangaroo and birds in a Wedge-tailed Eagle nest.

Figure 8.16 Searching for prey remains at a male eagle's roost. Nest and young are in the background.

to analyse orts and ignore pellets. Often, it depends on what the researcher has at hand, rather than which makes the better analysis.

A comprehensive coverage of the problems and contradictions around recommendations for prey analysis is not appropriate here, but it is important to note that Bonelli's Eagles differ from many other eagles around the world in that almost no orts are found under their nests, even when young are nearly full-grown (J. Real and L. Palma, pers. comm.). They seem to cart orts away.

Wedge-tailed Eagle or Little Eagle nests at this stage usually have plenty of castings and orts in the nest and under it. Sharp *et al.* (2002a,b) did not collect castings and orts from Wedge-tailed Eagle nests containing young or search under roosts used by adults. They found old nests, mostly not used in that season, and dismantled these nests searching for pellets and orts among the sticks. This method is quite different from most others used to study eagles' diets. Sharp *et al.* (2002b) concluded that combining pellets and orts for diet analysis might bias representations of prey species because of double counting. However, using only pellets or only orts limits the analysis. Certain prey species can occur in orts but not in pellets, or in pellets but not orts, and Sharp *et al.* (2002b) found exactly this, but recommended the use of orts anyway.

Many researchers combine orts and castings, taking care not to double count. Seguin *et al.* (1998) compared observations of prey delivered to nests, orts and pellets at an active Golden Eagle nest on Corsica and recommended that researchers combine pellets and orts for analysing diet if direct observations cannot be made. To avoid double counting, they quantified anatomical parts, e.g. jaws, teeth, femurs, to determine a Minimum Number of Individual estimates (MNI) from pellets and orts.

For example, if the teeth of 10 European Rabbits are counted in the castings found under a Wedge-tailed Eagle roost in the Brindabella Ranges, and the legs of eight rabbits are found in the orts under the nearby nest of the same pair, the biologist could assume a Minimum Number of Individuals (MNI) of 10 rabbits, not 18. In contrast, the approach often used on many raptor species would count rabbit found in five different castings as five different rabbits. We don't do this in Australia because we have seen both adults share a single rabbit or piece of kangaroo with nestlings, so four castings containing rabbit or kangaroo could be from one prey animal.

The third set of evidence recommended by Seguin *et al.* and many other biologists comes from observations, watching a nest through binoculars or telescope, or setting up a camera to record prey deliveries. However, in most studies, cameras or long periods in a hide are impossible, and counting prey delivered to a nest does not necessarily mean raptors ate this prey. Males often 'overdo it' and bring in more prey than the family can eat. Furthermore, some raptors, especially males, eat prey away from the nest and observations at nests provide little data on the prey taken during the very long post-fledging periods when many raptors feed their young away from nests. So all three types of data – orts, castings and observations – give a better estimate of raptor diet than does any one dataset alone, and this is the method used for most prey estimates in this book.

With observations, orts and castings combined, biologists use two separate methods to estimate which prey species are most important to a raptor. In the first method, the biologist counts the number of bones or feathers, especially skulls, primaries (the large flight feathers), feet, and certain long bones, pieces of fur, or mammal tails, to get an estimate of the minimum *number* of a particular prey species found. For example, they might find orts and castings in a Wedge-tailed Eagle nest that gave evidence for 17 individuals – 10 skulls of **Australian Raven**, 5 skulls of **European Rabbit**, and 2 skulls of **Grey Kangaroo**. Counting this contribution of prey by number would give us:

Australian Raven × 10 = (58.8%)
European Rabbit × 5 = (29.4%)
Grey Kangaroo × 2 = (11.8%)

One would need larger samples to draw firm conclusions about the diet at this nest, but from this small collection a biologist might propose that Australian Raven, at 59%, is the most important prey by number, European Rabbit at 30% is second in importance, and Grey Kangaroo third.

A second way of estimating the importance of prey in a raptor's diet is *biomass*, the actual amount of food represented by a prey item, the weight of meat if you like. Biologists calculate biomass because one prey species weighs more than another and, consequently, over time a large prey species could represent a larger percentage of the eagle's diet than is represented by the number of times it appears in the diet. Because an **Australian Raven** weighs ~650 g, a **European Rabbit** weighs 2500 g, and a **Grey Kangaroo** weighs 22 750 g, a biologist could calculate the importance of each of these three prey species in the example above by multiplying the number of each by the species' weight:

Australian Raven: 650 g × 10 = 6500 g (10.1%)
European Rabbit: 2500 g × 5 = 12 500 g (19.4%)
Grey Kangaroo: 22 750 g × 2 = 45 500 g (70.5%)

This gives a different estimate of the importance of Australian Raven in the diet compared with European Rabbit or Grey Kangaroo. It also gives a total of 64 500 g from 17 prey items, and a mean (average) prey weight of 3794 g. Biologists use another factor called wastage to measure the percentage of a large animal like a kangaroo that predators waste and discard because they can't eat it all in one sitting, but we don't include wastage factors in our discussion here.

So, in the example above, they found evidence of only two Grey Kangaroos at this nest, but they were the most important prey by weight, or biomass (the case at most high country Wedge-tailed Eagles' nests). A larger sample can change these findings. In studies with larger sample sizes, the most important prey species in terms of biomass for Wedge-tailed Eagles could be European Rabbits or ravens. So, we include prey by number and by biomass in the diet estimates in each species account below. It is important to understand that assessments of raptor diets are essentially indices, not precise estimates. They can't be exact because there are too many variables, and individual raptors and raptor species differ widely in foraging behaviour and their treatment of prey at the nest and roost. These estimates provide standardised indices for comparing diets of different raptors at the same time, and across seasons and years.

Breeding and non-breeding diet

As mentioned previously, Geometric Prey Weight of Wedge-tailed Eagles was 1650 g in one study, 1298 g in another. Though Wedge-tailed Eagle males and females tend to be ~1 kg lighter than North American Golden Eagles, one study found mean prey weight for Golden Eagles to be 690 g. Why the difference? As also mentioned previously, Australia has no vultures and Wedge-tailed Eagles take at least some carrion during breeding, whereas Golden Eagles tend to take more carrion during the non-breeding season. We have no good studies of the non-breeding diet of Wedge-tailed Eagles, and their diet could differ from the breeding season. Most studies of raptor diet are done from orts and castings collected at nest sites which could be misleading if the diet is different at other times of the year. For example, Nankeen Kestrels tend to catch arthropods, mice or insects during much of the year but sometimes increase the number of birds in their diet, such as Australasian Pipits or Common Starlings, when feeding their young. This makes sense because the male can make fewer trips between the nest and hunting area if he captures a few large prey items instead of many small prey items. He conserves energy and there may be certain prey, like young vulnerable starlings, available during the breeding season but unavailable at other times. Compared to a just-fledged starling in summer, an experienced adult starling in spring would be difficult for even an experienced kestrel to catch. Only for the Brown Falcon, Nankeen Kestrel, Peregrine and Southern Boobook do we have data on seasonal changes in diet. Mooney (cited in Cade 1982) found that the summer diet of the Brown Falcon in Tasmania was ~40% insects, 40% reptiles, 10% birds and mammals and 10% carrion. In winter the diet shifted to 30% small mammals, 30% carrion, 20% birds and insects and 15% reptiles. Newly independent young Brown Falcons depended heavily on concentrations of insects. Olsen *et al.* (1993) found that the breeding and non-breeding diet of Peregrines did not differ appreciably. Peregrines in Australia appear to have evolved as hunters of Galahs and smaller parrots, but they now eat introduced Feral Pigeons and Common Starlings. Of 988 prey items collected from Peregrine nests near Canberra, at least 41 species of birds were found. Of the prey items 96% were birds – Galahs, Crimson Rosellas, Eastern Rosellas, Feral Pigeons and Common Starlings were the main prey and made up 90% of the diet by weight. These main elements in the Peregrine diet near Canberra were remarkably stable regardless of season or changes in the abundance of certain prey across seasons (Figure 8.17).

For Southern Boobooks in Canberra most prey items for breeding owls were invertebrates (60.9%), among which beetles (Coleoptera 39.3%) and Lepidoptera

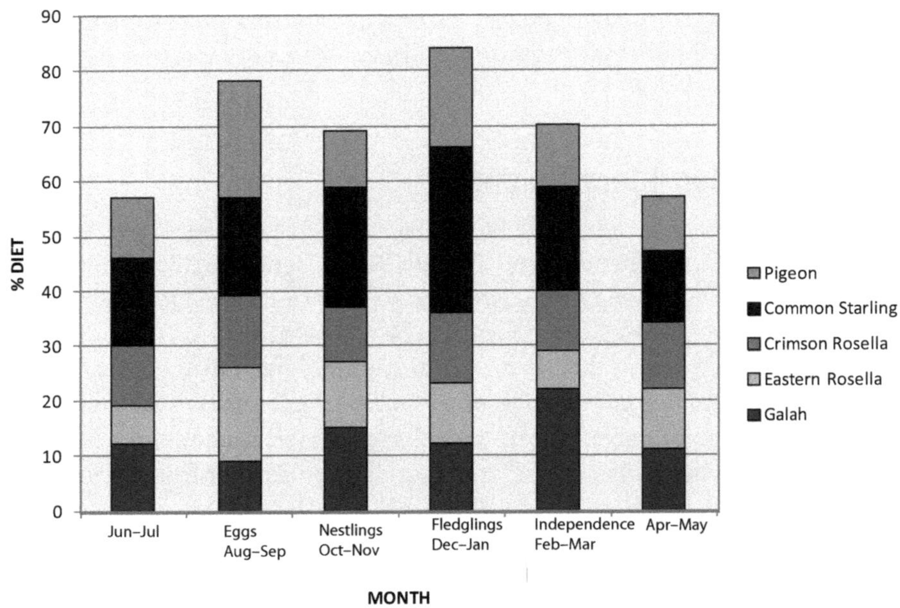

Figure 8.17 Peregrines near Canberra took the same prey in winter as during breeding (after Olsen *et al.* 1993).

(10.9%) were the most abundant. Birds, however, provided most of the dietary mass (89.4%), mainly Common Starlings (24.5%) and juvenile Crimson Rosellas (32.1%). In the non-breeding season just over half the Southern Boobook prey items by number were invertebrates (59.7%). Of the invertebrates, wolf spiders (14.7%), grasshoppers (9.5%) and butterflies and moths (8.9%) were the most abundant. However, vertebrates provided 97.2% of the total biomass. Mammals contributed 61.1%. The dominant mammal was the House Mouse (n = 154, 40.6% of biomass). Birds provided 36.1%, of which most were Common Starlings (17.1% of biomass). Invertebrates provided only 2.8% of the total biomass. There was a difference in diet between autumn and winter, with a higher proportion of vertebrates in winter (Table 8.2). In autumn the diet by number consisted of 25.4% vertebrates and 74.6% invertebrates, compared to 66.3% vertebrates and 33.7% invertebrates in winter ($\chi 2 = 78.22$; $P < 0.0001$).

Many raptors, including Peregrines, tend to take larger prey when breeding is successful than when breeding is unsuccessful, perhaps because they are bringing larger prey to their young (Olsen 1990).

Conspicuous prey

Raptors are known to select conspicuous prey, such as birds that look or behave abnormally. For example, they catch a white pigeon from a flock of dark birds.

Table 8.2 Comparison of autumn diet versus winter diet of Canberra Southern Boobooks, based on number of prey items from pellets and orts (after Trost *et al.* 2008)

	Autumn	Winter
Mammal	23.5% (*n* = 74)	51.4% (*n* = 93)
Bird	1.9% (*n* = 6)	14.9% (*n* = 27)
Reptile	0.0% (*n* = 0)	0.0% (*n* = 0)
% Total vertebrates	25.4% (*n* = 80)	66.3% (*n* = 120)
% Total invertebrates	74.6% (*n* = 235)	33.7% (*n* = 61)
Total prey items	*n* = 315	*n* = 181

Charles Darwin in *The Origin of Species* counselled pigeon-fanciers that 'hawks are guided by eyesight to their prey – so much so that on parts of the continent people are warned not to keep white pigeons, as being most liable to destruction'. It may not be colour itself but individuality in the flock that draws a hunting falcon to the 'odd bird out'. Derek Ratcliffe, the eminent British Peregrine expert, believed that this 'odd bird out' notion showed how Peregrines, as agents of natural selection, weed out substandard individuals in the prey population.

Each raptor eye has two foveas, acute focal points where receptors are especially numerous: a central fovea picks out movement, the temporal (side) fovea picks out detail. Raptors are experts at picking out newness (to the area) or defects in prey from a great distance, defects that make prey easier to catch. This is done with the temporal (side) foveas of both eyes (Figure 7.10).

Raptors constantly audit the landscape for an easy meal, for example weak or disabled prey, a limping rabbit, birds with missing feathers, young prey, or prey new to the environment, that is, prey that does not know the location of the raptor, or the local cover, the best ways to escape. In Canberra we commonly see raptors take conspicuous or odd prey. Australian Hobbies commonly take Budgerigars at a much higher rate than Canberra birdwatchers see them. These aviary escapees call, don't know hiding places, don't know where the Hobbies live and hunt, and they don't last long when a Hobby is around. Powerful Owls in Namadgi National Park took a Squirrel Glider *Petaurus norfolcensis*, a species said not to occur in the ACT. Peregrines at Bendora Dam in Namadgi National Park caught a Long-tailed Jaeger, an arctic-breeding, marine-wintering bird never previously seen in the ACT. Peregrines at the same site took a Black Falcon, a bird never reported in the Brindabellas. Little Eagles, Southern Boobooks and Collared Sparrowhawks in the ACT all took White-browed Babblers, said to be extinct in the ACT, and Wedge-tailed Eagles caused a stir when they took Long-nosed Bandicoots *Perameles nasuta* at two ACT eagle nests, marsupials said not to occur in the ACT. There are two messages here: 1) raptors are better at finding rare animals than we are; and 2) raptors may take animals newly released in an environment, for example, Brown Treecreepers or Southern Bettongs *Bettongia gaimardi* released at the fenced area at Mulligans Flat Nature Reserve in northern Canberra, as is planned. On paper, these two species are safe from raptors. In the 5000 or so prey items we have collected from ACT raptors, we have never found Brown Treecreepers, so they must number none or only a few, and Tasmanian Bettongs are largely nocturnal, but eagles can take them if they are flushed in the daytime or at dusk. Released Brown Treecreepers and bettongs could meet heavy predation, because they are new. 'Odd prey' theory trumps 'normal' foraging theory. Also, the key to finding some of these rare or unknown prey animals, such as bandicoots, is to radio-tag the adult eagles and follow them to their hunting grounds.

Food requirements

Food requirements mean how much food a raptor needs to stay alive, hunt, maintain a territory and fledge young. Kenward (1990) estimated that small hawks and falcons in captivity each day consume flesh weighing ~25–30% of their bodyweight. Large eagles consume ~5%. The amount of food consumed daily by a raptor relates not only to the size of the raptor but its hunting style – smaller quick-flying raptors such as sparrowhawks need proportionally more food than do larger slow-flapping raptors, so the Collared Sparrowhawk might require 25% of its weight per day to fuel its metabolic and energy needs whereas the Wedge-tailed Eagle might require only ~5%. For a large eagle, one week without food means little, but for a Sparrowhawk a week without food could be lethal. The greater surface area relative to body volume of a Sparrowhawk allows heat to escape more rapidly than it does from a Wedge-tailed Eagle (a slice of toast cools more rapidly than a freshly baked loaf of bread). In addition, smaller raptors that hover, such as kestrels or Black-shouldered Kites, spend proportionally more of the day actively foraging than do larger species like Wedge-tailed Eagles or Little Eagles, which often perch for much of the day. Mean food consumption of Australian owls hasn't been measured but in some North American owls it has – Great Horned Owl 4.7% of bodyweight, Burrowing Owls 15.9% of bodyweight, Pygmy Owl 50% of bodyweight (Peeters and Peeters 2005; Marti 1973; Johnsgard 2002).

Calculating food requirements involves several variables. Falconers have known for centuries that prey species vary in nutritional quality. They might feed their goshawk 100 g of rabbit or chicken, low-quality light meat, compared to only 50 g of hare or grouse, a high-quality red meat, for the goshawk to maintain the same bodyweight. In addition, a resting falcon, like an incubating Peregrine, uses far less food than an actively hunting falcon. Woodford (1960) gave average daily requirements for European Peregrines flown in falconry as 141 g for females and

113 g for males. European Peregrines are slightly larger than Australian Peregrines, females average 1125 g compared to ~960 g in the high country birds we have weighed; males average 680 g compared to ~660 g. For North American Peregrines, similar in size to European Peregrines, Craighead and Craighead (1956) calculated daily food requirements for three captives. The average food intake varied from 15% of bodyweight in autumn/winter to 11.5% in summer.

These, however, are resting metabolic values and flight is energetically very expensive, so wild raptors require much more food. Cade (1982) calculated the theoretical standard resting metabolic rate (SMR) for a 600 g male and 900 g female Peregrine based on a weight-relative equation derived for non-passerine birds – $SMR = 78.3 \times W^{.723}$ (Lasiewski and Dawson 1967). The female required 72.56 kilocalories per day, the male 54.12 kilocalories per day. Falcons probably used at least 10 times this much energy when flying at maximum speed. Cade then considered the energy required to carry prey back to the nest and known prey capture rates of hunting male and female Peregrines. Males were 93% successful on birds weighing ~90 g and females 31% successful on heavier grebes and coots. Because the female captured larger prey, she may offset lower success rates in hunting. To deliver 1000 kilocalories of food to the nest cost the male 49 kilocalories and the female 43 kilocalories.

In contrast, a 200 g Eurasian Hobby, similar to the Australasian Hobby, had a calculated SMR of 1.02 kilocalories per hour. Flying after termites weighing one-tenth of a gram each expends, perhaps, $10 \times SMR = 10.2$ kilocalorie per hour. It must catch and eat ~51 termites just to break even. To feed a brood of four young, the Hobby would, perhaps, have to catch an additional 200 termites per hour for five hours so it is clear why many falcons, including Hobbies, switch to birds and other large prey during the breeding season. It is energetically more efficient. It also explains why, paradoxically, the smaller males in species such as Southern Boobooks, several kestrel species, and kites such as Mississippi Kites, catch heavier prey on average than do their larger female mates. It is energetically more efficient to catch larger prey farther from the nest and cart it back to the nest, so males travel farther for the benefit of larger vertebrate prey, while females guard the eggs and young and catch available invertebrate prey around the nest where vertebrate prey is rare.

Tarboton (1978) measured energy use in Black-shouldered Kites *Elanus caeruleus* in Africa, which, he notes, are very similar in size and habits to the Australian Black-shouldered Kite. The African kites still-hunted from perches for 71% of their hunting time and by hovering 29% of the time. Hovering had a 6.9 times higher energy rate than perch (still) hunting, but hovering produced 2.5 times the strike opportunities and 2.2 times higher strike success rate. The observed daily energy intake by Black-shouldered Kites was 99.7 kcal. Hunting from a hovering position accounted for 30% of its day but 61% of its daily energy use.

Table 8.3 Basic diet of high country raptors that can be summarised as F, I, B, R, M, C

Key: F=Fish, I=Insects, B=Birds, R=Reptiles, M=Mammals, C=Carrion

Species	Main prey
Black-shouldered Kite	M
Whistling Kite	FBMCI
White-bellied Sea-Eagle	FBR
Swamp Harrier	MCB
Brown Goshawk	IBM
Collared Sparrowhawk	IB
Grey Goshawk	IBM
Wedge-tailed Eagle	MRBC
Little Eagle	MRB
Brown Falcon	IBRMC
Australian Hobby	IB
Peregrine Falcon	B
Nankeen Kestrel	IBRM
Eastern Barn Owl	M
Sooty Owl	M
Masked Owl	M
Powerful Owl	MB
Southern Boobook	MBIR

Basic breeding diets

As mentioned above, Australian high country raptors differ from their Northern Hemisphere counterparts in that Australian raptors: 1) use more carrion during breeding; 2) use more invertebrates during breeding; and 3) do not have native grassland rodents as breeding prey (we have no diurnal ground rodents in the high country). However, Australian raptors have adapted to several introduced bird and mammal species.

The basic diets for raptors breeding in the Australian high country are summarised in Table 8.3.

A histogram of recorded prey for each diurnal raptor species in the high country, where data are available, is given in each chapter in Species Accounts.

Fast-flying species such as Peregrines (Figure 8.18) and Hobbies tend to catch more birds than do more buoyant Brown Falcons, kestrels, harriers, eagles and kites, which take more prey from the ground. For example, Baker-Gabb (1984) showed that nesting Peregrines and Black Falcons near Mildura, Victoria, caught ~18% Galahs by number and both captured a large percentage of Common Starlings. However, the Black Falcon made up the remainder of its diet with rabbits whereas the Peregrine shunned rabbits, even though they were particularly numerous, making up the rest of its diet with other birds. Brown Falcons, slower and more buoyant than Peregrine or Black Falcons, ate more rabbits than the other two falcons. Gause's Rule tells us that coexisting species must differ in their feeding ecology. There is some controversy about Gause's Rule, but it seems to hold for the feeding ecology in sympatric Australian falcons and large eagles.

Figure 8.18 Collecting prey remains from a Peregrine nest at Ginninderra Falls.

9

Eagles in the high country

The first major scientific study of any Australian raptor described the diet of Wedge-tailed Eagles in the high country, and serious raptor research in Australia begins with this study – Leopold and Wolfe (1970). One purpose of the study was to understand the levels of lamb killing by Wedge-tails which at the time were heavily persecuted in Australia. Until 1968, two states, Queensland and Western Australia, paid bounties on eagle scalps, averaging between them payments on 13 056 scalps per year for the period 1958–67. Western Australia had terminated bounty payments, but in 1970 the species was still unprotected in New South Wales, Victoria, South Australia and the Northern Territory. Only Tasmania and the Australian Capital Territory protected them. The data were fragmentary, said Leopold and Wolfe, but the total kill of Wedge-tailed Eagles in Australia could exceed 30 000 birds per year. They asked: was this killing justified?

Undoubtedly, eagles were capable of killing lambs and occasionally did so. But there were no reliable data showing that eagles were a financial drain on the sheep industry. Most evidence of lamb predation was fragmentary or anecdotal, not quantitative. The CSIRO Division of Wildlife Research, based at Gungahlin in Canberra, supported their study and much of it was conducted in the northern portion of the Australian Capital Territory and the adjacent New South Wales tablelands, between Canberra and Lake George, in an area typical of much of the sheep range in south-eastern Australia. For comparison, they gathered limited data from some eagle nests at lower elevations in more arid areas in north-western New South Wales and south-western Queensland.

To contextualise the 'eagle problem', Leopold and Wolfe summarised the history of the sheep industry in the south-eastern high country. They used reports from Southey (1848), Westgarth (1853) and Walker (1963). The livestock industry in Australia started with settlement in 1788 but gained momentum after 1835 with the importation of additional flocks of sheep, cattle and horses. At that time the highlands of southern New South Wales were heavily timbered, though valleys were kept more or less open by Aboriginal burning practices. Sheep were herded in flocks of 500 or less and were kept moving by the shepherd and his assistant or 'hut-keeper'. Because of guarding shepherds, one would assume that eagles took fewer lambs compared to flocks without shepherds.

As forested land was cleared by ringbarking (or girdling) trees, and pastures were further improved by introducing high-yield grasses, a shortage of labour forced the abandonment of shepherding and the confinement of sheep to fenced paddocks.

Our colleagues in Spain, Marcos Moleón and Iñigo Zuberogoitia, relate a similar history of change away from shepherding to fencing there, and the commensurate increased persecution of Golden Eagles. During the period of development of pastoralism in Australia, another Spanish animal became of great relevance to the welfare of Wedge-tailed Eagles. The European Rabbit was introduced near Geelong, Victoria, in 1859, and quickly became a major scourge over the rangelands. Other eagle prey mammals introduced to the high country included the European Hare, Red Fox and various forms of domestic livestock.

In the early years of Australian sheep farming, the eagle was seldom mentioned as a pest, either because flocks were guarded by shepherds, or because there were fewer sheep. But from ~1860 on there are many published references to eagles, and many of them described eagles as a menace to lambing ewes, though an almost equal number extolled them as rabbit predators. Leopold and Wolfe cited two papers from the journal *Emu* as examples:

> *McGilp (*Emu *1925, p. 120): 'In the Lake Frome district (S.A.) as many as 300 eagles have been destroyed, and I saw so many alive that I gave orders that more attention to poisoning should be given …*
>
> *'On two occasions I saw eagles attacking stray lambs that but for my arrival would have been killed by the eagles …*
>
> *'I observed five eagles diving down on to a lamb that they were driving from its mother. The eagles took it in turn to dive and strike with the claws at the lamb. I rode up and found the lamb so torn about the shoulders, neck and back that I decided to destroy it and poison the carcass. Result, 11 dead eagles …*
>
> *'From past experience I know it is impossible to keep eagles and lambs …*
>
> *'Rabbits are fairly scarce, previous droughts and constant destruction by poison having almost cleared them out.'*

> *Lansell (*Emu *1936, p. 246): '… Wedge-tailed eagles (in the Riverina of New South Wales) very seldom kill a lamb and practically never kill a sheep over two months old. On each of two occasions when, to my knowledge, eagles killed a lamb, the lamb was either sick or one of twins and was deserted by its mother. Rabbits form the principal diet of eagles here and very many are killed during the year. As usual the eagles generally work in pairs, one disturbing the rabbit from its seat while the other catches it in its talons …*
>
> *'Having visited the nests of at least ten pairs for each of the last four years and never having found any trace of a lamb in or under the nests, it is quite fair to say that eagles, in this district, do far more good than harm. As many as 30 rabbit skulls have been in one nest and on several occasions fresh rabbits were found in the nest long before the eggs were hatched.'*

This conflict in beliefs about the role of Wedge-tailed Eagles in the pastoral industry could only be resolved by systematic research. Starker Leopold knew California Golden Eagles, birds very similar in size and habits to Wedge-tailed Eagles. But crunching over dry eucalypt leaves in sheep paddocks outside Canberra batting at flies was different. In February 1964 Leopold and Wolfe started checking nests, at a time of year when there was no breeding activity. In summer heat they gathered orts from nests known to have been active in the previous season. They noted, 'In addition to loose bones, feathers, tufts of hair, and miscellaneous animal parts, there were usually some pellets that had been regurgitated by the birds.' Between February and July, they gathered and identified these food remains from 31 unoccupied (because it was the non-breeding season) nest sites yielding 1923 identifiable food fragments and 157 pellets. When the 1964 nesting season began, Tony Wolfe located 18 occupied eagle nests near Canberra, and visited these every few days, collecting fresh prey items brought to the nests to feed the young. They identified 260 prey animals.

To compare the prey data collected around Canberra with that taken by eagles in more arid, lowland country, they searched an area in the arid interior of north-western New South Wales, north of White Cliffs, and in southern Queensland, near Cunnamulla. They found 28 eagle nests that yielded 1274 recognisable food fragments and 52 pellets.

They concluded that, by number, European Rabbits constituted the primary prey of eagles near Canberra. Eurasian Hares were the next most important species, followed by birds, sheep and lambs, small mammals, and lizards. In the arid interior rabbits were also important, but kangaroos and lizards were more important than hares, which didn't occur there. Lambs constituted ~7% of the diet. An unknown portion of this 7%, they said, was probably carrion (Table 9.1).

These observations make sense as there were probably more kangaroos and reptiles available in the warmer inland than around Canberra. This is an

Table 9.1 Minimum number of items in the diet of Wedge-tailed Eagles in the highlands around Canberra compared to lowlands in inland New South Wales/Queensland (Leopold and Wolfe 1970)

	High country		Inland low country		
Prey groups	*n*	%	*n*	%	**Weight (kg)**
Rabbit, *Oryctolagus cuniculus*	114	43.8	113	30.6	2.47
Hare, *Lepus capensis*	41	15.8			4.00
Macropods (kangaroos and wallabies)	5	1.9	94	25.5	19.50
Sheep, *Ovis aries*	4	1.5	3	0.8	40.00
Lamb	23	8.8	25	6.8	15.00
Other mammals	21	8.1	13	3.5	4.40
Crows, Ravens, *Corvus* sp.	14	5.4	18	4.9	0.61
Australian Magpie, *Craticus tibicen*	8	3.1	8	2.1	0.33
Parrots, Order Psittaciformes	9	3.5	15	4.1	0.36
Other birds	12	4.6	15	4.1	0.48
Reptiles	9	3.5	65	17.6	0.23
Totals	260	100	369	100	

important point; available prey around 1960 could have been different from available prey in 1980 or 2000.

To establish a figure for density of nesting pairs in the high country, they selected an area of ~250 square miles (648 km^2) north and east of Canberra and attempted to locate every active pair of resident eagles. They found 20 active pairs, at a density of about one pair per 12 square miles (31 km^2). Sheep flocks within an average eagle nesting territory produced ~4000 lambs, of which the pair of eagles might take at most a few individuals. Although the impact of predation by nesting eagles on the pastoral industry in the area could not have been significant, non-territorial wandering young eagles (floaters) might have caused trouble in some lambing paddocks, but they collected no data from these non-breeders. As a result of forest clearing and the introduction of rabbits and other exotic mammals, Leopold and Wolfe said that south-eastern Australia was now probably a better habitat for Wedge-tailed Eagles than it was originally. Figure 9.1 shows the location of nesting territories they found in their study.

Forty years after Leopold and Wolfe

Esteban Fuentes, a Masters student from Mexico, was pleased when he saw Starker Leopold's name on the Wedge-tailed Eagle paper. Esteban wanted to study Wedge-tailed Eagles in the high country and Starker Leopold was famous in Mexico for his books *Wildlife of Mexico* and *The California Quail*, as well as being an expert on predators. To compare Wedge-tailed Eagle diet in the same area some 40 years

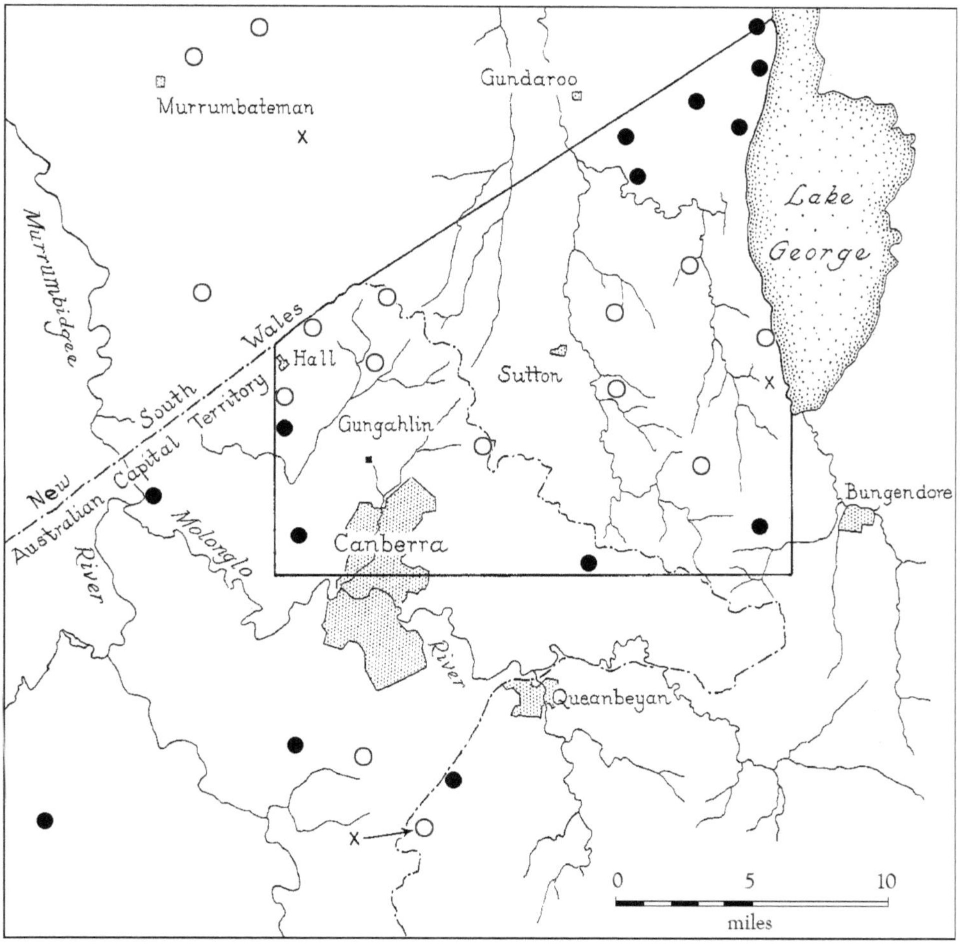

Figure 9.1 Location of Wedge-tailed Eagle nests found by *Leopold and Wolfe (1970)* and sampled for prey. Black dots = successful nests; white dots = failed or unknown result.

after the Leopold and Wolfe study we searched for occupied territories between July 2002 and May 2004, and extended the search into the Murrumbidgee River and Namadgi National Park. During the period of our study, the region was under severe drought.

Nesting and territory occupancy

During 2002–2003, we located 44 Wedge-tailed Eagle territories. We checked 26 of the 32 territories found by Leopold and Wolfe, 22 (85%) of which were still occupied. We located an additional 15 territories inside their study area that either they had missed or were newly occupied since 1964. Another seven territories were outside the Leopold and Wolfe study area in the mountainous southern part of the

Table 9.2 Reproductive success and productivity of Wedge-tailed Eagles on the Southern Tablelands in 1964* and 2002–2003**

	1964	2002–03
Total territories ***	15	**44**
% breeding *(eggs laid)*	86.7	93.9
% successful	66.7	83.7
No. of fledged young	12	56
Young fledged per territory	0.8 ± 0.2	1.1 ± 0.1
Young fledged per *active nest*	0.9 ± 0.2	1.2 ± 0.1
Young fledged per successful territory	1.2 ± 0.1	1.4 ± 0.1

* Taken from Leopold and Wolfe (1970, p. 14).
** Taken from Fuentes *et al.* 2007
*** Territories for which reproductive outcome was known.

Table 9.3 Percentage of Wedge-tailed Eagle territories fledging 0, 1 or 2 young in 1964 and 2002–2003 (from Fuentes *et al.* 2007)

No. young fledged	1964	2002–03
0	33.3%	16.3%
1	53.3%	53.1%
2	13.3%	30.6%

ACT. Contrary to their 1964 survey, nine nests were located inside the Canberra city limits, with an average distance to paved roads of 720 ± 132 m (range 130–1270 m) and to suburbs of 1117 ± 251 (range 260–2000 m). Four of those were less than 500 m from houses. Only one territory was completely surrounded by houses, but some eagles seemed to be adapting to humans after the years of heavy shooting, trapping and poisoning ended.

Breeding success and productivity

The reproductive success and productivity of the Wedge-tailed Eagles in both time periods is shown in Table 9.2. Compared with 1964, adults fledged more young per territory in 2002–2003, mainly because they fledged more broods of two young instead of one, and there were fewer pairs that failed to fledge any young at all (Table 9.3).

Comparing breeding diet: 1964 v. 2002–2003

In 2002–2003 we collected 492 prey items from 33 territories, and from castings, prey remains and observations we identified 57 different species: 19 mammals, 20 birds, 7 reptiles and 1 crustacean. Mammals dominated by biomass (95.3% of total dietary mass). Though rabbits dominated the 1964 diet, macropods dominated the 2002–2003 diet, representing 19.9% of the total items and 45.6% of the dietary mass

Figure 9.2 Wedge-tailed Eagles on the Southern Tablelands take many Eastern Grey Kangaroos including adults, mobile juveniles, and joeys ejected from the pouch when eagles chase adult females.

Table 9.4 Number, percentage contribution by number, and weights of different prey species and groups recorded in the diet of Wedge-tailed Eagles in and near the ACT, 1964 and 2002–2003, based on prey remains (Fuentes *et al.* 2007)

	1964		2002–03		
Prey groups	*n*	%	n	%	**Weight (kg)**
Rabbit, *Oryctolagus cuniculus*	114	43.8	56	17.0	2.47
Hare, *Lepus capensis*	41	15.8	26	7.9	4.00
Macropods (kangaroos and wallabies)	5	1.9	64	19.4	13.30
Sheep, *Ovis aries*	4	1.5	9	2.7	40.00
Lamb	23	8.8	6	1.8	15.00
Other mammals	21	8.1	22	6.7	4.40
Ravens, *Corvus* sp.	14	5.4	25	6.7	0.61
Australian Magpie, *Craticus tibicen*	8	3.1	25	6.7	0.33
Parrots, Order Psittaciformes	9	3.5	36	10.9	0.36
Other birds	12	4.6	56	18.8	0.48
Reptiles	9	3.5	5	1.5	0.23

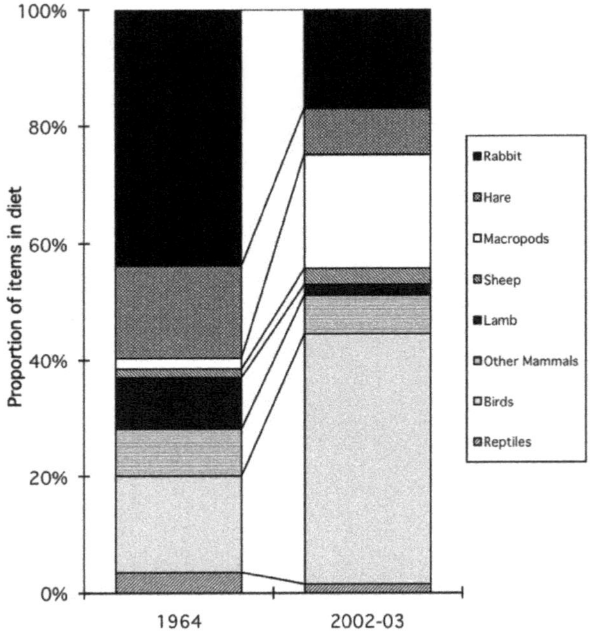

Figure 9.3 Proportional contribution of different prey groups to the diet of the Wedge-tailed Eagle by number, based on fresh prey remains collected during the breeding season on the Southern Tablelands in 1964 and 2002–2003 (from Fuentes *et al.* 2007).

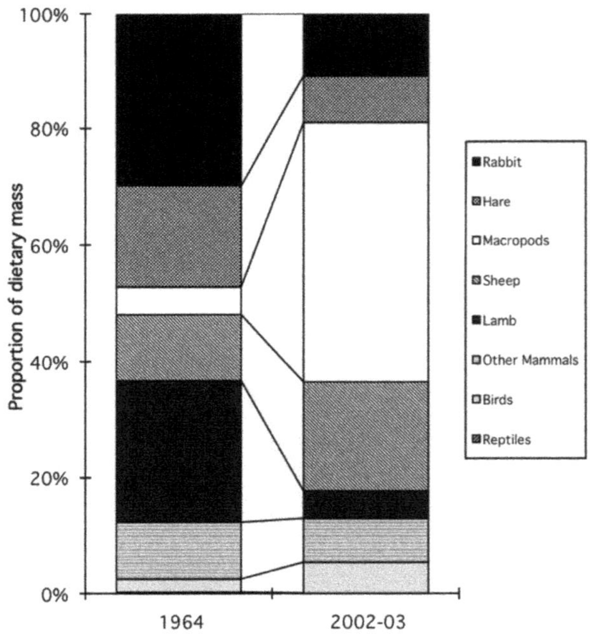

Figure 9.4 Proportional contribution of different prey groups to the dietary biomass of the Wedge-tailed Eagle, based on fresh prey remains collected during the breeding season on the Southern Tablelands in 1964 and 2002–2003 (from Fuentes *et al.* 2007).

Figure 9.5 Eastern Grey Kangaroo killed and eaten by Wedge-tailed Eagles (Peter Magassy).

(Figure 9.2). The most important prey was the Eastern Grey Kangaroo, *Macropus giganteus* (13.6% and 31.2% by number and biomass respectively). Other important items were rabbits and adult sheep. Among birds, the Galah and Australian Magpie were the most important bird prey species.

The proportional contribution of different prey categories was significantly different between the two time periods. By 2002–2003 three prey groups had *decreased* significantly in the eagle's diet: rabbit (from 43.8% to 17.0%); hare (from 15.8% to 7.9%); and lamb (8.9% *to* 1.8%). By 2002–2003 three other prey groups had significantly *increased*: macropods (1.9% to 19.4%); parrots (3.5% to 10.9%); and other birds (4.6% to 17%) (Table 9.4, Figures 9.3 and 9.4).

Biomass figures showed similar trends, with a marked decrease in the biomass contribution of rabbit, hare and lamb and an increase in the contribution of macropods (Figure 9.3).

In other words, there was a shift from a diet dominated by exotic animals such as rabbits and hares, to a more equal distribution of prey species. In 1964, exotics provided 72.1% of the prey items by number and 83% by biomass, while native animals contributed only 27.9% by number and 4.9% by biomass (12.1% of the items had an unknown origin). In 2002–2003 only 36.7% of the prey items by number and 42.5% by biomass were exotics and 62.4% by number and 44.6% by biomass were native animals (12.9% were unknown). These changes in the diet could be a response to changes in the prey populations. Eastern Grey Kangaroos (Figure 9.5) are said to have increased inside the city and in reserves outside the city such as Namadgi National Park and Googoong Foreshores (Don Fletcher and Environment ACT, unpublished data; Reardon 2003). The same applies to some ground-feeding birds such as Galahs, Sulphur-crested Cockatoos, ravens and magpies, all of which increased in the last 21 years (Veerman 2002). These shifts in prey numbers fit with the proportional changes we found in the most commonly taken prey. The concept of 'prey switching' is surprisingly controversial. The suggestion is that more abundant prey, such as rabbits, are taken by eagles at a disproportionally high rate compared to rarer prey, such as kangaroos. But if the relative abundance of the rarer prey species increases, proportionally more of these are taken until they are over-represented in the diet, whereas the initially favoured prey, rabbits in this case, are under-represented (Murdoch and Oaten 1975). So the proportional increase in kangaroos in the diet in this study area from 1964 compared to 2002–2003 could be related to a relative decrease in rabbits, compared to kangaroos, as well as a relative increase in kangaroos. Sharp *et al.* (2002a,b) reported that Wedge-tailed Eagles in western New South Wales may have switched from a diet dominated by rabbits to a diet that had proportionally more reptiles. The Norwegian raptor biologist Y. Hagen (1969) described similar processes. He classified *primary* prey species such as voles and hares, and *secondary* (scarcer) prey species such as grouse. Hagen showed that when mammal prey crashed, raptors switched to prey species such as grouse. Predator numbers and their breeding success remained stable in the absence of their primary mammal prey. Only after grouse also declined did raptors start to decline, and then they eventually switched back to increasing mammalian prey. In the Australian context it is important to note that European Rabbits are nutritionally inferior, white meat compared to the red meat of hares or kangaroos. A shift to predominantly rabbits is not necessarily a good thing for eagles. During drought, one would expect both rabbits and native prey such as macropods to decline, so drought reduces the breeding success of eagles. However, rabbit control, such as Rabbit Haemorrhagic Calicivirus Disease (RHD), introduced to cause rabbit declines, can result in an increase in native animals including macropods (Sandell and Start 1999) so we should consider the heretical notion that reducing rabbit numbers in non-drought years could *increase* the breeding success of raptors such as eagles or Barn Owls that rely on numerous and more nutritious macropods and/or other native mammals for breeding.

Shifts in the other direction, from native to introduced prey, have been noted for other high country raptors over a different time scale since European settlement, for example, Powerful and Sooty Owls (Bilney *et al.* 2006). But, probably, Powerful Owls and Sooty Owls had much less of a prey shift than any other raptors in the high country because other species such as eagles, Black-shouldered Kites, Peregrines, Brown Falcons and Hobbies take proportionally more introduced prey such as European Rabbits, House Mice, Black Rats, Rock Doves and European Starlings than do Powerful and Sooty Owls. Also, Aboriginal burning before European settlement could mean that some current Sooty Owl caves were once in more open country and, therefore, were actually Masked Owl caves. Boobooks and Barn Owls also use some of the same caves that Sooty Owls use and may have contributed to prey remains found there and used for these pre- and post-European settlement comparisons. In Chapter 10 we discuss changes in diet over an 11-year period for Peregrines.

Rabbits as prey in the high country

In the high country, higher average rainfall and a mosaic of different habitats, among other things, make the region less prone to extreme fluctuations in prey populations than Australia's interior. Furthermore, because of increased numbers of macropods, the Wedge-tailed Eagles in the 2002–2003 Canberra study had a broader prey base that could have increased their reproductive success compared with 1964. To us it seemed that as long as Wedge-tailed Eagles had alternative prey available, especially kangaroos, they could breed. Even when rabbits decrease, eagles can switch to other prey, and they may even prefer macropods over rabbits where macropods are abundant.

The role of alternative native prey species in eagle ecology is probably underestimated. It is widely believed that breeding numbers and breeding success of Wedge-tailed Eagles are linked to rabbit abundance, as in the theoretical shown in Figure 9.6.

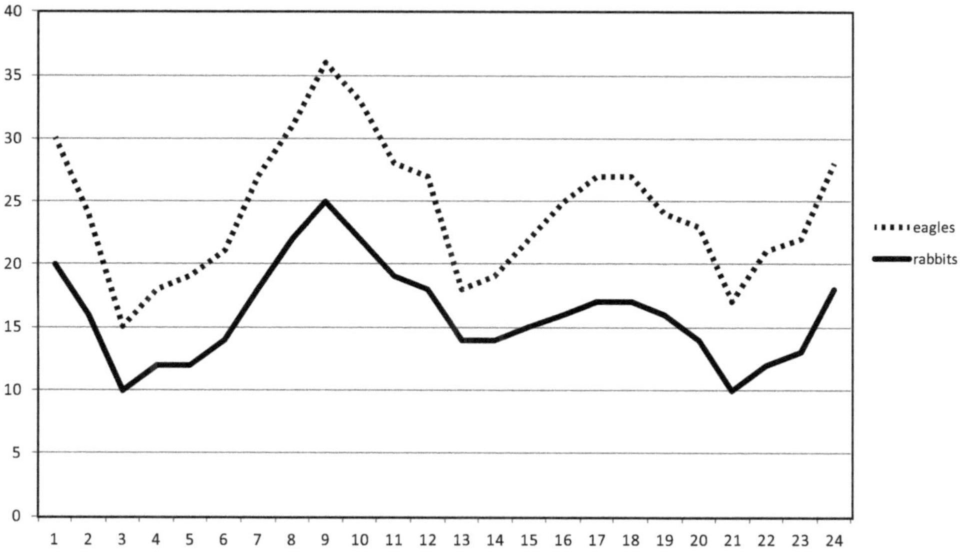

Figure 9.6 Theoretical link between rabbit numbers and eagle numbers or breeding success.

The belief that Wedge-tail and Little Eagle numbers and their breeding success are tied to rabbit numbers is a popular story underpinning much of the scientific literature, a story too powerful to challenge. So let's challenge it.

Is the breeding success of Wedge-tailed Eagles and Little Eagles influenced by European Rabbit control?

Little Eagles had been declining in the ACT for about a decade and the usual suspects were trotted out, lack of food, particularly dwindling European Rabbits. This principle, declining rabbits = declining raptors, was chiselled in stone. It was written into the nomination for listing the Little Eagle as Vulnerable in the ACT (COG 2008). Rabbit Haemorrhagic Calicivirus Disease (RHD) was said to be a likely cause of rabbit decline. There were two problems with this notion: 1) lack of precedent: there were no published data showing that densities of any breeding raptor species in non-drought Australia increased or decreased in relation to rabbit numbers. Steele and Baker-Gabb (2009) compared raptor abundance before and after the introduction of RHD in 1995–96, as a pest-control mechanism that caused severe declines in rabbit numbers. They found no detectable effect from the introduction of RHD on populations of raptors said to be dependent on rabbits as food. Still, the story persisted and was tied to declining Little Eagle numbers. 2) Though rabbits in some parts of Australia had been decreasing, rabbits in the ACT were *increasing*. As rabbits increased (Figure 9.7), Wedge-tailed Eagles and Little Eagles included proportionally more of them in their diets (Table 9.5) while breeding Little Eagle numbers fell.

RHD escaped from Wardang Island, off South Australia, in October 1995, and may have been widespread in the ACT by late 1996 (Brian Cooke pers. comm.). Studies following the arrival of RHD on the Australian mainland showed that the virus caused mortality rates of up to 95% in the rabbit populations investigated, but was less effective in cooler regions of Australia. Henzell *et al.* (2002) described a cline in the effectiveness of RHD along a hot-dry to cool-humid gradient, and Saunders *et al.* (1999) showed that the impact of RHD in the New South Wales Central Tablelands was patchy. Presumably, a similar situation applied in the south-eastern high country where the rabbit death rate was far lower than in the dry country inland. Antibodies against RHD were found in sera of high country

Table 9.5 Proportion of rabbits by number in the diet of Wedge-tailed Eagles and Little Eagles near Canberra during three periods, 2002–2003, 2004–2007 and 2008–2009, as rabbit availability increased (J. Olsen unpublished data)

	2002–2003	2004–2007	2008–2009
Wedge-tailed Eagle	16.6%	26.8%	34.0%
Little Eagle	23.6%	41.9%	50.0%

Figure 9.7 Male Wedge-tailed Eagles delivered many rabbits to the nest as rabbit numbers increased (and these were quantified in diet statistics), but often these were left lying around the nest uneaten.

Figure 9.8 Adult Whistling Kite found dead under its nest in an area of Pindone use.

rabbits sampled before the introduction of RHD. These two observations, combined, led to the hypothesis that a similar benign virus had already been present in Australian wild rabbits, giving them partial immunity. This virus was discovered, isolated and its genome published (Strive *et al.* 2010) and termed Rabbit Calicivirus Australia 1 (RCV-A1).

To counter the decreased effects of biological rabbit control in the high country, including Canberra, the chemicals Pindone (2-pivalyl-1, 3-indandione) and 1080 (sodium fluoroacetate) were commonly used to control rabbits. In the 1970s and 1980s researchers such as John McElroy thoroughly tested the effects of 1080 on native animals. Effects of this chemical on wildlife were tested and known, but Pindone had not been extensively tested. In fact, one government biologist told me that the world centre for intensive Pindone use was the New South Wales highlands, around Bowral. Pindone was accepted for use in Australia with almost no toxicity trials or research, and most other countries do not use it. The US Environmental Protection Agency (EPA) evaluated Pindone but decided not to register it for use in the United States. At high doses, Pindone is fatal to raptors, or disables them temporarily (Martin *et al.* 1994), which can be fatal if the raptor can't forage or evade predators. If rabbit control measures were affecting raptors in the high country, via secondary poisoning, relatively more Little Eagles than Wedge-tailed Eagles could die, because Little Eagles took proportionally more rabbits (by biomass 52% for Little Eagles versus 13% for Wedge-tailed Eagles in one study (Olsen *et al.* 2010), 73% versus 19% in a more recent study (Olsen *et al.* 2013b): Table 9.6). Little Eagles may be more sensitive to such poisons than other raptors, though we are waiting for hard evidence to support this suggestion.

Little Eagle and Wedge-tailed Eagle diets were essentially different. Geometric Mean Prey Weight for Wedge-tailed Eagle prey in the more recent study was 1650 g, for Little Eagles 337 g, similar to the 5-fold difference in weight between male Little Eagles and male Wedge-tailed Eagles. The great difference in GMPW and heavy use of rabbits by Little Eagles suggest that there is little interspecific competition for prey.

Table 9.6 Summary of dietary parameters of Wedge-tailed Eagles (WtE) and Little Eagles (LE) breeding sympatrically in the ACT, 2008–2009

	% n		% biomass	
Diet	**WtE**	**LE**	**WtE**	**LE**
Mammals	65	52	92.6	72.8
Rabbit: adult	31	16	19.4	35.5
Rabbit: juvenile	3	35	0.7	36.2
Birds	33	45	7.2	23.6
Reptiles	1	3	0.2	3.4
Dietary statistics	**WtE**		**LE**	
GMPW	1650		336.7	

To test the belief that Little Eagles were declining because of declining rabbits, we compared the numbers of nesting Little Eagles in the ACT against numbers of rabbits. Data on numbers of rabbits/km came from similar transect counts to those used in studies of Wedge-tailed Eagles by Ridpath and Brooker (1986) and Parker *et al.* (2007) (discussed below). The method used gave a trend for increase or decrease in numbers of rabbits, rather than a population density.

Figure 9.9 shows there was no significant relationship between the decline in numbers of active Little Eagle nests and trends in numbers of rabbits, even after the arrival of RHD by 1996. Rabbit numbers peaked around 2006–2007 to the point of requiring active control, and peaked again in 2010, whereas the number of breeding pairs of Little Eagles had declined to less than 10% of what it was in the early 1990s. In recent years (2007–2010) there may even be a negative relationship between the Little Eagle breeding status and rabbit numbers, and in 2011 there was only one breeding pair of Little Eagles found in the ACT (Olsen *et al.* 2012). There was no sign of a lagged recovery of the Little Eagle breeding population following the resurgence of rabbits (Figures 9.9 and 9.10).

The lack of a correlation between Little Eagle breeding numbers in the ACT and trends in rabbit numbers suggested that some reason other than RHD and loss of

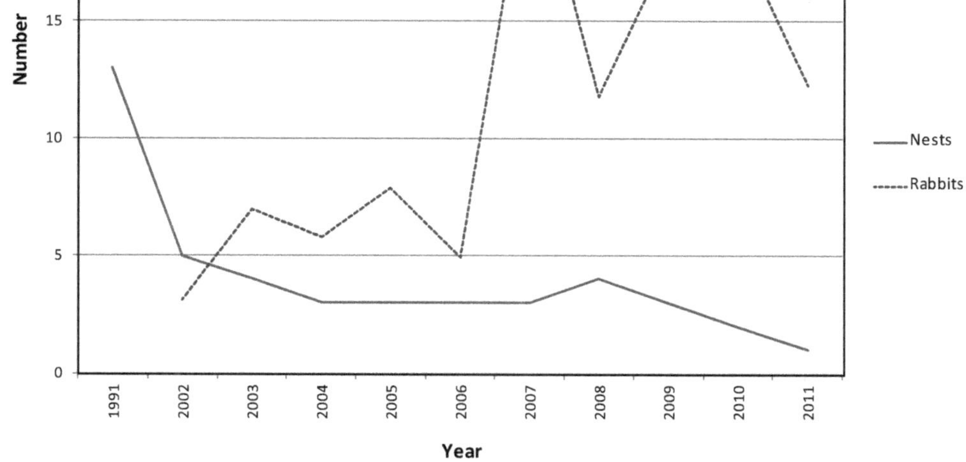

Figure 9.9 Relationship between number of active nests of Little Eagles (from annual surveys of breeding pair, = Nests) and abundance index of rabbits (rabbits/km, from monthly surveys, see text, = Rabbits) in the ACT, 1990–2011 (regression: $F = 1.98$; $P = 0.20$; not significant).

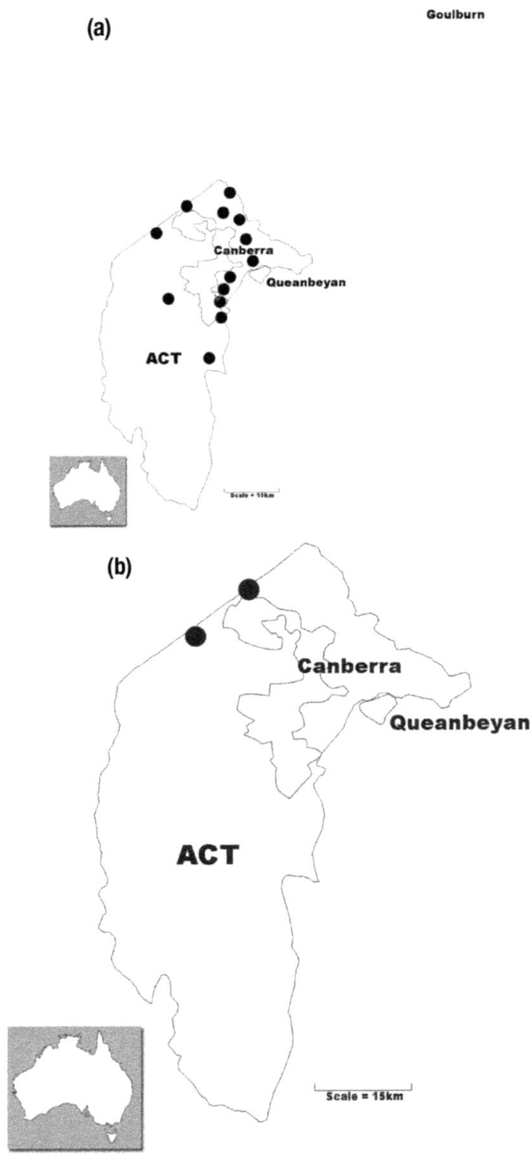

Figure 9.10 Breeding Little Eagles found in the Australian Capital Territory: (a) 1992, (b) 2011. Dots represent clusters of nests.

prey was causing the Little Eagle's decline. Various possible reasons had been canvassed, for example, increasing competition with Wedge-tailed Eagles for decreasing habitat and nest sites (owing to urban expansion), and poisoning (e.g. Olsen *et al.* 2009, 2013a). We had previously refuted the idea that Wedge-tailed Eagles competed substantially with Little Eagles for prey because Little Eagles took

Figure 9.11 This female Wedge-tailed Eagle became progressively ill and finally died on her eggs.

more rabbits than did Wedge-tailed Eagles and more rabbit kittens as opposed to adults, but Wedge-tailed Eagles could compete for nest sites. 'Floater' Little Eagles continued to appear in the ACT, maybe from the wider continental population, and we wondered if the ACT region had become a 'population sink' for this species.

What about Pindone? Pindone was and still is used mainly in peri-urban areas (G. Saunders pers. comm.) right across the high country, because it is much less toxic to dogs than is 1080, and an antidote exists, whereas there is no antidote for 1080. Adult Little Eagles, Wedge-tailed Eagles and Whistling Kites had been found, inexplicably dead, in or under their respective nests, resulting in nest failure, but these bodies were not analysed for Pindone (Figure 9.11).

At the time we recommended that: 1) the Little Eagle be uplisted from *vulnerable* to *endangered* in the ACT (based on a >90% decline in two generations which easily meets IUCN criteria for the *endangered* category); and 2) Pindone be banned in the remaining Little Eagle home range(s) with occupied nests, as a precautionary measure, pending the outcome of investigations into Pindone use and any correlations with defunct Little Eagle territories.

As mentioned above, Pindone was accepted for use in Australia with almost no toxicity trials or research (see Martin *et al.* 1994) and used extensively across the

Figure 9.12 David Mallinson climbing to a Little Eagle nest in the 1980s.

high country, so the problem could be more widespread than was realised. It could also bear on the 50% decline in atlas reporting rates for the Little Eagle in south-eastern NSW over the past 30 years (see Canberra Ornithologists Group 2008). It would be ironic if Pindone, which appears not to be controlling rabbit populations, was killing off a predator in the high country that eats mostly rabbits.

Evidence for Australian eagle dependence on rabbits

While we were gathering evidence and ideas about reasons for the Little Eagle decline (Figure 9.12), scientists were still saying to us, 'It must be RHD causing rabbit declines and, therefore, causing the Little Eagle decline.' This mystified us. Rabbits in the high country were in plague proportions, you saw them everywhere; a half million dollars was set aside by the ACT government to battle rabbits. The story of eagle dependence on rabbits was more convincing than the evidence, or lack of it. How had this powerful story evolved? Often the link between rabbits and eagles was made with no evidence, for example, Shephard *et al.* (2005) suggested that European Rabbits may have assisted the increase of inland populations of White-bellied Sea-Eagles, but there is no evidence for this claim and Olsen *et al.* (2006c) found no evidence that rabbits were an important component in the diet of White-bellied Sea-Eagles in the high country. Evidence for eagle–rabbit interactions in Australia are drawn mainly from analogies to overseas systems. Several species of small rodent (voles and lemmings) undergo regular population fluctuations of 4–5-year cycles in North America and Fennoscandia, while in Northern Canada Snowshoe Hares *Lepus americanus* show similar cyclic patterns in ~10-year cycles. Raptors in the Northern Hemisphere respond to increasing or decreasing lagomorph and rodent numbers in at least two ways: *numerical response*, raptors increase or decrease in number, or *breeding response*, more or fewer pairs breed, produce more or fewer eggs and/or young. People assume that Australian raptors cycle with introduced rabbits in the same way, even though rabbits are different from hares, in a different genus with different life histories. Regardless, Australian researchers use these Northern Hemisphere studies as analogue templates, and assume that drought in Australia, and/or diseases such as *myxomatosis* act on rabbit numbers, and on Australian raptors, in the same way that cycling hares act on raptors in the Northern Hemisphere. Below, I examine the validity of the evidence collected in early Australian studies on eagle–rabbit interactions, particularly with Wedge-tailed Eagles, the source of most research and most claims about eagle dependence on rabbits.

The history of research on Wedge-tailed Eagles in Australia has two broad phases. *Phase 1* – early researchers emphasised the reliance of Wedge-tailed Eagles on the European Rabbit, in part to de-emphasise the eagle's economic impact on sheep, and also to fit the prevailing view that Wedge-tailed Eagles were rabbit predators. This research speculated about how *myxomatosis* affected eagle breeding; the studies include Leopold and Wolfe (1970) summarised in the chapter above, as well as Baker-Gabb (1984), Ridpath and Brooker (1986), Robertson (1987) and Olsen and Marples (1992). *Phase 2 – myxomatosis* was used as an analogy to predict cause-and-effect changes resulting from Rabbit Haemorrhagic Calicivirus Disease (RHD). Introduced around 1996–97, researchers predicted that RHD would reduce rabbit numbers (which it generally did), then reduce eagle numbers and breeding success. However, evidence for declines in eagle numbers or breeding success did not eventuate (Edwards *et al.* 2002a,b; Steele and Baker-Gabb 2009). How solid was the original evidence that Wedge-tailed Eagle numbers and breeding success were affected after *myxomatosis* was introduced in early 1950? Researchers predicting the effects of RHD on eagles based these predictions largely on three studies: 1) the proportion of rabbit in the diet of different raptor species in Victoria (Baker-Gabb 1984); 2) a threshold minimum number of 1.6 rabbits/kilometre, determined through transect counts, needed for Wedge-tailed Eagles to breed (Ridpath and Brooker (1986); and 3) clutch-size reductions in different raptor species reported by Olsen and Marples (1992) after the *myxomatosis* virus was introduced in Australia.

Measuring eagle breeding success

In the beginning, Leopold and Wolfe (1970) said that rabbits were the most important prey of Wedge-tailed Eagles in the high country around Canberra, but kangaroos were more important out west in the arid interior. They did not say that rabbits were essential for breeding or that Wedge-tailed Eagle numbers fluctuated with rabbit numbers, and they provided the first comprehensive measure of breeding success for the species. To compare the breeding success found in these early studies to breeding success found in later studies it is important to standardise these figures and understand the various indices used. Some studies used *young per clutch laid* (i.e. if only one pair in five laid eggs, and the pair produced two young, then *young per clutch* = 2), or *young per successful pair* (also called *brood size*, i.e. if only one pair in five fledges young, and the pair has two young, *young per successful pair* = 2). However, in a long-lived, K-selected species such as the Wedge-tailed Eagle, I find that *young per territory per year* is the best measure (i.e. if only one pair in five fledges young, and the pair has two young, then *young per territory per year* = 0.20). This last measure links breeding to food availability, weather and other variables, and assumes that Wedge-tailed Eagle pair densities remain fairly constant over many years, which we know they do. As noted above, Fuentes *et al.* (2007) found slightly higher productivity than did Leopold and Wolfe – 1.1 versus 0.8 *young per successful pair*, for Wedge-tailed Eagles in the same area some 40 years later. Table 9.7 shows *young per territory per year* from several studies and the proportion of rabbits and macropods by number in the diet.

Table 9.7 Wedge-tailed Eagle breeding success and percentage of rabbits and macropods in diet by number

1 – Leopold and Wolfe (1970), 2 – Ridpath and Brooker (1986) a) unstocked b) cattle c) Dry West Coast d) sheep e) Mediterranean, 3 – Robertson (1987), 4 – Fuentes *et al.* (2007), 5 – Baker-Gabb (1984), 6 – N. Mooney, pers. comm.., (*Tasmanian* study), 7 – Parker *et al.* (2007), 8 – Silva and Croft (2007), 9 – Collins and Croft (2007)

Study	1	2a	2b	2c	2d	2e	3	4	5	6	7	8	9
Young/terr/year	0.8	0.19	0.19	0.37	0.46	1.02	0.99	1.10	1.00	0.64	1.20	0.60	0.64
% no. rabbit	43.8	97.5	83.6	9.60	70.6	33.5	55.0	16.9	74.7	8.0	10.2	33.0	5.0
% no. wallaby and kangaroo	1.9	0.7	3.0	17.7	5.9	2.7	16.0	19.4	–	45.0	21.74	6.0	20.0

Phase 1 Wedge-tailed Eagle research

1 – **Ridpath and Brooker (1986)** recorded the lowest productivity (young per territory per year) of any study, 0.19 *young per territory per year* in two areas, 2a and 2b (Table 9.7). However, three other areas in their study, 2c, 2d, 2e, where pairs took fewer rabbits, showed higher productivity. They did not find a relationship between high rabbit numbers and Wedge-tailed Eagle numbers or breeding performance in any of the areas, so they established a threshold for eagle breeding of *1.6 rabbits/kilometre*, less than which the eagles could not breed, based on the two areas of marginal country where eagle productivity was poor. This threshold has been cited as a baseline in several other studies that also measured rabbit numbers, but none of these studies could link the 1.6 rabbits/kilometre threshold to breeding productivity (see below). Ridpath and Brooker were able to link decreasing eagle breeding success in these two marginal areas to low overall rainfall: drought depressed rabbit numbers and eagle breeding performance, but the authors did not quantify the eagle's other main prey such as macropods, which probably also declined during drought, leaving no alternative prey. Anecdotally, they stated that macropods declined, but only rabbits got attention. In other words, the authors structured this baseline around the two areas because overall productivity was low, and they didn't use the other three study areas because eagle productivity was too high and regular, and the proportion of rabbits in the diet too low. The authors didn't discuss the apparent advantages to breeding eagles of having proportionally less rabbit in the diet.

2 – From 1980 to 1985 **Robertson (1987)** examined the effect of drought on a breeding population of Wedge-tailed Eagles near Menindee, in western New South Wales. The number of eagles successfully breeding varied from 10–12 pairs during favourable years to three pairs during drought. He counted rabbits on transects similar to those used by Ridpath and Brooker (1986). Outbreaks of *myxomatosis* in all years affected rabbit numbers but, overall, the number of breeding eagles was not significantly correlated with the density of rabbits, except at fledging, well after eggs were laid or nestlings hatched, and after nestlings were being fed. The index of

abundance of rabbits at Kinchega fell below the 1.6 rabbits/km threshold, but eagles still bred.

Near Menindee, breeding success averaged 0.99 chicks per pair (range 0.73–1.25) and remained high when food was scarce, during the drought, and during *myxomatosis* outbreaks. About 14 500 kangaroos died at Kinchega during the drought of 1982 (Robertson 1987). The high breeding success of Wedge-tailed Eagles breeding in 1982 when live prey was scarce, he said, may have been either an artefact of the small sample size or an indication that pairs that bred were experienced breeders and efficient at hunting. The high breeding success of eagles breeding in 1983, when the density of rabbits on transects was low, according to Robertson, may reflect the completion of pouch life of the first post-drought cohort of kangaroos, suitable prey for adults feeding large chicks.

3 – **Baker-Gabb (1984)** made the first comparative dietary study of Australian raptors (12 species). Of the seven rabbit-specialists, the Brown Falcon, Brown Goshawk, Black Kite and Little Eagle had biomass of rabbit above 85% in their diets, and Wedge-tailed Eagles, Spotted Harriers and Whistling Kites had biomass of rabbit less than 85%. Of the rabbit-specialists, the Wedge-tailed Eagle had the least rabbit by biomass, 69.6% (see Table 9.8).

These comparative findings of per cent biomass in raptor diets were broadly supported by later studies conducted south of 25° latitude in Australia where rabbits were most common. Aumann (1988) studied Brown Goshawks in Victoria and found that they took 52.1% rabbits by biomass; McDonald *et al.* (2003) studied Brown Falcons in Victoria and found that they took 61.8% rabbits by biomass; Olsen *et al.* (2010a) compared the diets of Wedge-tailed Eagles and Little Eagles breeding near Canberra and found that Little Eagles took 52.4% rabbits by biomass compared to 12.7% for Wedge-tailed Eagles, a lower overall proportion of rabbit than Baker-Gabb found, but reflecting the difference between the two eagles – Little Eagles take proportionally more rabbits than do Wedge-tailed Eagles.

4 – **Olsen and Marples (1992)** studied the clutch sizes of 24 diurnal Australian raptor species from eggs collected mainly by amateurs before 1954 and after 1953

Table 9.8 (i) Per cent biomass in diet of main rabbit predators near Mildura, Victoria (after Baker-Gabb 1984); and (ii) decrease in clutch post-1953 (after Olsen and Marples 1992). % biomass rabbit in the diet above 85% and significant clutch size reduction are ***bold italicised***

	Wedge-tailed Eagle	Spotted Harrier	Whistling Kite	Brown Falcon	Brown Goshawk	Black Kite	Little Eagle
(i) % rabbit in biomass	69.6	80.0	83.1	***87.0***	***89.8***	***91.3***	***92.1***
(ii) clutch size decrease post-1953	***Yes***	No	***Yes***	No	***Yes***	No	No

when *myxomatosis* reduced rabbit numbers, mainly below 25° latitude where rabbits were most abundant. They found a mean decrease in clutch size in only three of 24 species, Brown Goshawks, Whistling Kites and Wedge-tailed Eagles, and this has fed the existing narrative that rabbit numbers strongly affect the breeding numbers and success of Wedge-tailed Eagles. Aspects of this study could form questions for further study. Other factors occurring after 1953 that could influence clutch sizes included factors such as weather, different egg collectors, laws against egg collecting, areas where *myxomatosis* hit versus areas where it did not, numbers of alternative main prey such as kangaroos, and reduced persecution of Wedge-tailed Eagles after 1953. Leopold and Wolfe (1970) said that the number of Wedge-tailed Eagles killed *each year* in Australia could exceed 30 000. The bounty on Wedge-tailed Eagles was lifted in Western Australia around 1968, and until 1968, 13 056 scalps per year were handed in. The species was first protected in South Australia in 1972, in New South Wales in 1974 and in Victoria in 1975, so there were many more eagles competing for limited food after the 1970s than in the 1950s and 1960s when persecution was high. Could dividing the samples pre- and post-1975 instead of pre- and post-1953 better correlate with differences in clutch sizes? Baker-Gabb (1984) who studied prey taken by breeding raptors in western Victoria (Table 9.8), showed that the species that took the most rabbit, by biomass, were Black Kite, Little Eagle, Brown Goshawk, and Brown Falcon, but Olsen and Marples (1992) found no decline in clutch size after 1953 for any but the Brown Goshawk. In their sample, the clutch size of Black Kites actually *increased*. In short, among the seven rabbit predators studied by Baker-Gabb, the findings of clutch depletions in three species by Olsen and Marples seem almost random (Table 9.8). In Table 9.9 below we gathered clutch-size data for Wedge-tailed Eagles

Table 9.9 Clutch-size in Wedge-tailed Eagles pre (1954) and post (1953) the main impact of *myxomatosis* (introduced 1950).

For this study, all available eggs from WA Museum, Museum of Victoria, Australian National Wildlife Collection, Australian Museum and SA Museum. Pre = pre-*myxomatosis* 1859 to and including 1953. Post = post-*myxomatosis* 1954–1979 (number of clutches in brackets). 1952–1959 is the period just after *myxomatosis* was introduced (1950) and rabbit numbers initially fell by some 99%. *1990–1995 is a period when rabbit numbers were said to be relatively high just before the introduction of RHD in 1995.*

Period	This study	Olsen and Marples 1992
pre-1954	1.93 (222)	1.98 (282)
post-(1954–1979)	1.84 (343)	1.84 (184)
1950–1959	1.98 (40)	
1952–1959	2.000 (39)	
1960–1969	1.81 (101)	
1970–1979	1.82 (101)	
1980–1989	1.85 (74)	
1990–1995	*1.80 (35)*	

from five museums, compared these data to the Olsen and Marples data, but also split the sample into blocks of years.

As you can see, pre- and post-1954 data in the two samples are similar, but clutch size in our sample actually *increased* in the 1950s just after *myxomatosis* was introduced, then decreased in the 1960s when rabbits were said to have begun increasing again, particularly in the 1990s just before RHD was introduced.

The claim that Wedge-tailed Eagle clutch size declined after 1953 could be a myth, though one that received currency and unquestioning support.

Phase 2 Wedge-tailed Eagle research

Newsome *et al.* (1997) assessed the likely impacts on wildlife, including raptors, when rabbits declined from RHD. Their assessments were largely speculative but, based on the previous studies, concluded without further evidence that Wedge-tailed Eagles and Little Eagles were the species most likely to decline post-RHD and experience the most severe impact on breeding.

Sandell and Start (1999) noted that, so far, no correlations had been established between the breeding activity of raptors and rabbit numbers. Indeed they noted that counts of raptors done between 1996 when RHD was introduced and March 1998 showed that that Wedge-tailed Eagles numbers had *increased* in south-western and south-eastern Australia. However, they reasoned that eagles may have been drawn to roads, where these counts were done, looking for road kill after rabbits had declined. They provided no evidence to support this conjecture. They noted that native herbivores, particularly large kangaroos, could increase in the wake of reduced rabbit numbers, but they did not say that increased kangaroos might benefit breeding Wedge-tailed Eagles.

From May 1996 over a 2.5-year period **Edwards *et al.* (2002a,b)** measured rabbit numbers before and after extensive ripping of burrow systems and the added effect of RHD. Rabbits declined by 85% but numbers of Wedge-tailed Eagles and other raptors did not. The authors were surprised at this, they said, because rabbit declines had a demonstrable effect on the breeding activity of Wedge-tailed Eagles, according to Ridpath and Brooker (1986) and Olsen and Marples (1992). They reasoned 'That we failed to show a population reduction for raptors may be due to the low population densities of raptors in arid areas and the fact that our fixed transects were small relative to the size of areas used by some species, particularly Wedge-tailed Eagles' citing Ridpath and Brooker (1986).

In contrast, some researchers have studied breeding Wedge-tailed Eagles where rabbits were scarce or absent. **Collins and Croft (2007)** studied breeding Wedge-tailed Eagles in western New South Wales; at five nests the eagles produced six clutches of two eggs, fledged a mean 0.64 young, and took mostly macropods. **Winkel (2007)** studied Wedge-tailed Eagle breeding in north Queensland where there are no rabbits and found that they successfully bred by taking prey such as large kangaroos. In south-central Queensland around Charleville **Parker *et al.* (2007)** tested the threshold of 1.6 rabbits/kilometre suggested for successful breeding of Wedge-tailed Eagles by Ridpath and Brooker (1986). Rabbit densities were too low to detect any rabbits by this spotlighting method. From 13 active nests at five sites (four sheep stations, one national park) over one breeding season they found nine clutches, all of two eggs, and a mean of 1.2 fledglings raised per attempt, at the high end of the recorded range for Wedge-tailed Eagle productivity. By biomass, the eagles' diet was dominated by kangaroos, especially Euros, and Eastern Grey and Red Kangaroos, with few rabbits taken among a range of mammals, birds and lizards (total 28 species). They also took other mammals such as cats, foxes, pigs, goats and lambs, some birds, notably Emus and Bustards, and some dragon lizards. They concluded that in northern Australia the diversity and abundance of small and juvenile macropods, rather than rabbits, may influence breeding success.

In summary, the claim that eagle numbers and breeding success are linked to rabbit numbers could be a myth, in the high country and elsewhere. Eagles probably don't need rabbits to breed, as long as other prey is available. During drought, one would expect both rabbits and native prey such as macropods to decline, so drought reduces the breeding success of eagles.

Powerful Owls roost on prey during the day (J. Olsen).

In addition to hovering, Black-shouldered Kites perch and watch for prey (Julian Robinson).

Male Masked Owl (Rohan Bilney).

Male Sooty Owls are much smaller than their female mates (Rohan Bilney).

Barn Owls frequently occur in the high country but haven't been found breeding there (Rohan Bilney).

Female Brown Goshawks perch in a tree close to the nest to guard nestlings (J. Olsen).

Nestling Peregrines defend themselves against enemies, including humans (Michael Maconachie).

Placing a just-fledged Wedge-tailed Eagle in a tree close to its nest (Chapter 14) (Sue Trost).

Juvenile White-bellied Sea-Eagles have bare tarsi, unlike juvenile Wedge-tailed Eagles (Derek Willcox).

In flight, White-bellied Sea-Eagles have a distinctive wing shape and short tail (Rohan Bilney).

Black-shouldered Kites study the ground looking for rodents (Julian Robinson).

Brown Goshawk returning to its nest area with prey (Julian Robinson).

Aerial food pass from male to female Black-shouldered Kites (Charles Davis).

Peregrines hunt above the tree canopy (Mark Osgood).

Just fledged Nankeen Kestrels wait for their parents to deliver food (Charles Davis).

Nestling Wedge-tailed Eagles in the high country (Charles Davis).

Unlike species such as Peregrines, Wedge-tailed Eagles have very flexible wingtips (Charles Davis).

Adult White-bellied Sea-Eagle attacking a juvenile Sea-Eagle (Julian Robinson).

Whistling Kites have a distinctive underwing pattern (Julian Robinson).

Grey Goshawks stray into the high country but they haven't been found breeding there (David Pastern).

Nestling Peregrine Falcons brought up from their cliff nest for banding (Sue Trost).

Nankeen Kestrels prefer ground prey such as rodents (Roger Williams).

Collared Sparrowhawks prefer to hunt birds below the tree canopy (Roger Williams).

Little Eagles have different colour-phases (Julian Robinson).

Black-shouldered Kites can fledge two broods a year (Julian Robinson).

As with other diurnal raptors, Nankeen Kestrels are as heavy as their parents just before fledging (Julian Robinson).

Brown Falcons have a distinctive double malar stripe on their faces, and a 'pot-bellied' look when perched (Julian Robinson).

Brown Falcons have short toes for running along the ground, and for catching reptiles. (Julian Robinson).

Wedge-tailed eagles are capable of powerful, swift flight (Julian Robinson).

Brown Falcons have rounder wingtips than other falcons (Julian Robinson).

10

Peregrines in the high country

Wedge-tailed Eagles and Peregrine Falcons are the most studied diurnal raptors in Australia. Both are iconic species but there are other reasons for their receiving so much attention. As mentioned above, Wedge-tails were seen as lamb killers, then seen as indicators of the effects on rabbits of *myxomatosis* and Rabbit Haemorrhagic Disease, so were heavily studied. By contrast Peregrines, which occur worldwide, were the first species linked to the phenomenon of eggshell thinning and the pesticide DDT. Peregrines specialise in killing flock bird species such as domesticated homing pigeons, Rock Doves, which is the main reason Peregrines were intensively studied at first. The two aspects are intertwined and the story begins in Great Britain.

Derek Ratcliffe and DDT

A day in the English countryside with Derek Ratcliffe was a wonderful experience. He showed us two Grass Snakes *Natrix natrix*, Britain's largest reptile, an old Sparrowhawk nest, deer, and other aspects of nature in the Kings Forest near Thetford; he then detailed facts and stories about each of them. At one point he gave me a leg up to a Tawny Owl nest in an oak and warned me to guard my eyes; the adults could attack. He was kind and generous with his time but often turned dark when the subject drifted to government policy or big business, his opponents, the wreckers and parasites of Britain.

Figure 10.1 The author with Derek Ratcliffe in southern England in 1999.

For many scientists, Derek Ratcliffe was peerless. On his sudden and unexpected death at the age of 75, Ian Newton wrote that Britain had lost the most outstanding field naturalist and conservationist of his generation. Derek (Figures 10.1, 10.2), received awards from around the world but not from the State. Contemporaries who had contributed much less to conservation and cutting-edge science received nods from the Queen but Derek was ignored. Some thought he stood up to big government too effectively and too often, especially when he challenged the powerful forestry lobby in Britain in the 1980s. He effectively persuaded the government to halt the tax-breaks that drove forest planting of the wild Flow Country in northern Scotland. Forestry interests wanted to plant trees over open ground, and Derek successfully challenged this. He knew the value of these open upland habitats because he had roamed through them and studied them throughout his life. He was a concise and outspoken civil servant, the

Figure 10.2 Derek Ratcliffe watching a wombat in Tidbinbilla Nature Reserve.

architect of the 1977 Nature Conservation Review that identified the most important wildlife habitats in Britain. By the time he visited Canberra in the late 1980s, Derek had heavily influenced the study of raptors in Australia. Today we take for granted the link between agricultural pesticides and declines in birds of prey, but Derek was the first person to make this link and harden the link with evidence. He discovered that organochlorines killed Peregrines and thinned their eggshells. The discovery spurred raptor studies around the world, especially those involving Peregrine Falcons.

The story of Derek's discovery began in wartime England. Peregrines had always been abundant there, a favourite of medieval falconers. You could even visit nest ledges still occupied in the 20th century that were known to have been occupied by Peregrines in the time of Henry VIII, a keen falconer. Peregrines were protected in the early 20th century, but the protection was dropped during World War II by *The Destruction of Peregrine Falcons Order* dated 1 July 1940 by the Secretary of State for Air. The British military put written messages into small capsules attached to homing pigeons that accompanied personnel on various missions. Downed airmen on sea patrol off the western and north-eastern coasts of Britain wrote their location on these messages, released the homing pigeon and waited for rescue. Peregrines, as pigeon-killers, caught some of these, so the thinking was that these falcons needed to be eliminated from Britain. The military, accompanied by zealous citizens, shot adults, smashed the eggs on cliffs, destroyed nestlings, and significantly reduced the number of breeding pairs, especially in Cornwall and Devon. *The Destruction of Peregrine Falcons Order* expired in February 1946 and the protection of Peregrines and their eggs was restored, though some gamekeepers still shot and trapped birds that interfered with the gentlemanly sport of game shooting. Peregrines quickly recovered to pre-1940 levels.

However, the Peregrine recovery didn't make pigeon-fanciers happy, and they agitated to remove them from the protected list. This militancy was fierce in the 1960s in mining valleys in South Wales, the long-time stronghold of pigeon-fanciers. Ivor George from Neath, their spokesman, appeared on television arguing that Peregrine numbers were increasing and the falcons were slaughtering loved and valued pigeons. A petition followed and the Home Office, through the Nature Conservancy, launched a nationwide survey of Peregrines, with Derek Ratcliffe as organiser.

It is important to note that this tension between pigeon-fanciers and the falcons was just as strong in Australia, even into the 1990s where Peregrines were illegally killed, especially in Tasmania. In the mountain eyries in Namadgi National Park we commonly found racing pigeon bands (Figure 10.3). Apparently the fanciers drove their pigeons to Canberra and released them so they could fly to Victoria. The winner was the one that arrived home in the quickest time. Peregrines took many of these pigeons as they flew through mountain passes, and many pigeon-fanciers publicly threatened retribution.

Figure 10.3 Racing pigeon bands from Namadgi National Park Peregrine nests.

In 1960s England, Derek organised a network of people to survey nesting Peregrines and investigate claims of damage to the pigeon racing sport, but found, to his dismay, that Peregrine numbers were in a headlong crash, not increasing. Nobody knew why. Scientists didn't understand the dark side of pesticides in 1960; these chemicals were seen as miracle compounds with low toxicity to humans, effective in controlling insects that spread diseases such as malaria. Rachel Carson had yet to publish *Silent Spring* (1962). Scientists in 1960 wondered about diseases similar to *myxomatosis* that had recently decimated British rabbits. Could Peregrines be crashing because of a similar disease? Other researchers wondered about cyclodienes, the organochlorine group of chemicals including dieldrin, aldrin and heptachlor, used as seed dressings. Eventually they found that these were toxic and did kill songbirds, but there was no clear connection between the deaths of seed-eating birds on farmland and Peregrines in the distant hill country. The catastrophic decline continued. The other strange thing was that Derek noticed Peregrines eating their own eggs. Why would they do this? He eventually worked out that dieldrin, aldrin and heptachlor might kill Peregrines outright, but another organochlorine, DDT, was thinning Peregrine eggs and the thinned eggs crushed under the weight of the incubating adult who then ate the contents. But he needed more evidence. He knew that egg collecting had been a popular British hobby for decades. By comparing the weight of eggs from old collections with the addled eggs he found in nests, Derek found that the shells of modern eggs had grown substantially thinner and more prone to cracking. Moreover, they contained residues of agricultural pesticides. He went to museums and weighed eggs taken by collectors before the introduction of DDT after World War II. Some newer eggs seemed lighter in weight than older eggs but this was confounded by eggs of different sizes and different volumes. Derek pondered this and eventually developed an Eggshell Thickness Index, often called the Ratcliffe Index, calculated this way:

Weight of the eggshell / length × breadth of the eggshell

The Ratcliffe Index changed everything. Because egg collectors had operated in so many countries, researchers in North America, Eurasia and Australia could measure the thickness of Peregrine eggs before and after the introduction of DDT in the 1940s and 1950s in their country. And also, because Peregrines use the same nesting cliffs for many generations, researchers could plot the occupancy of cliff nests and the productivity (young per territory or young per pair) of Peregrines against DDT use.

Declines were found in many parts of the world, including certain parts of Australia. Different countries around the world reacted to the Peregrine decline in different ways. DDT was banned in the UK in 1984. The European Union banned it in 1986, the USA in 1973 and Australia banned it in 1987. However, the chemical

Figure 10.4 Spanish Peregrines like this one were imported into the United States by biologists and released to form hybrids.

was still manufactured and sold overseas, for example, in South America. In Britain and Scandinavia Peregrines recovered on their own. In the United States Peregrines had disappeared east of the Mississippi River but still bred in certain parts of the west, like Arizona, and in Alaska. Researchers decided to breed Peregrines in captivity and release these, but, controversially, took the small subspecies *Falco peregrinus brookei* from Spain (Figure 10.4), where Peregrine populations were high because DDT use was low. Americans even imported some *F. p. macropus* from Australia, bred these and released them in the United States, intentionally or accidentally lost by falconers, to form hybrids with the recovering *F. p. anatum*. Falconers could buy Australian Peregrines in the United States in those years, mostly bred from the young of a single pair in Melbourne Zoo. The wisdom of this genetic mixing has been challenged, and DNA analysis of Peregrines now breeding in places like New Jersey show that they are largely Spanish *brookei*, not the American *anatum* subspecies (Michael Wink, pers. comm.).

Peregrines are still among the most widely and best studied birds in the world. Following Ratcliffe's lead, in 1973 we started an investigation of Peregrine numbers and pesticides in the Flinders Ranges of South Australia, and in 1975 Ken Norris, Bill Emison and Bill Brenn started a statewide survey of Peregrines in

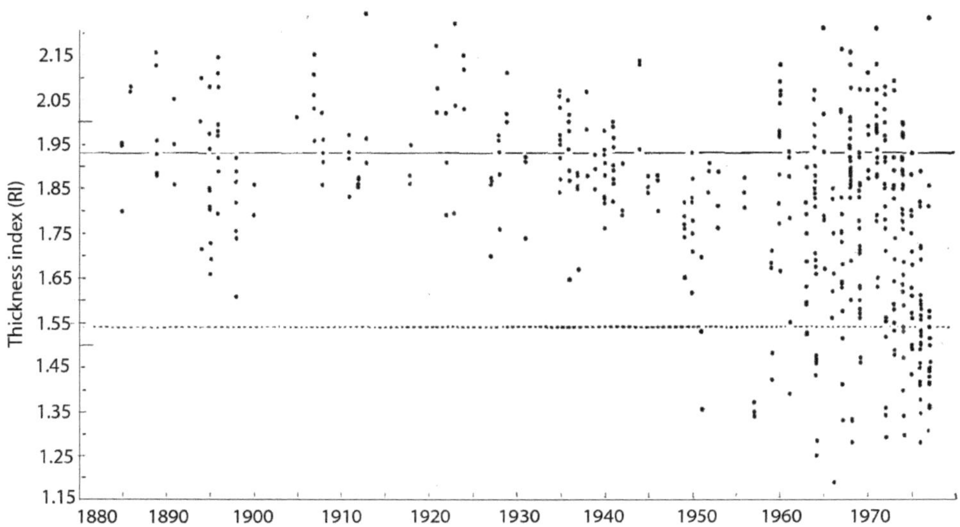

Figure 10.5 Eggshell thinning in Australian Peregrine eggs. Top line is the mean, bottom line is 20% thinner than normal (after Olsen and Olsen 1979).

Victoria (Norris *et al.* 1977). No population of North American raptor had been able to maintain a stable, self-perpetuating population with a decrease in eggshell thickness of 18% or more; Derek Ratcliffe said that a 19% decrease in shell thickness was associated with egg-breakage in the British Peregrine. The only pollutant consistently linked with shell thinning was DDE, a metabolite of DDT. DDT was introduced into Australia around 1947 and like other countries, Australia had many egg collectors so pre-DDT eggs could be compared to post-DDT eggs. As you can see in Figure 10.5 there was considerable thinning after 1947.

All states sampled showed some thinning, but Victoria showed the most (Figure 10.6), though none reached the 19% threshold statewide that spelt the demise of the Peregrine. Occasionally Peregrines are found dead (Figure 10.7).

Peregrines in the Australian high country were virtually unaffected by DDT. As in the United States and Canada, Australian Peregrines were bred in captivity and released into the countryside around Canberra, but this was experimental, not an effort to bolster populations that had never fallen in the first place. Pigeon-fanciers got wind of these releases and immediately complained that the government was breeding up Peregrines and releasing them near Canberra to hurt their sport. As with so many myths about raptors, this was nonsense.

It is important to remember that DDT did not affect all raptors in equal measure. In North America, Peregrines were nearly wiped out but the similar-sized Prairie Falcon was not, so some Prairie Falcons moved into abandoned Peregrine nest sites. Of the two North American eagles, Bald Eagles were nearly wiped out in

Figure 10.6 Thinned Peregrine eggs in Victoria crushed by the female's weight.

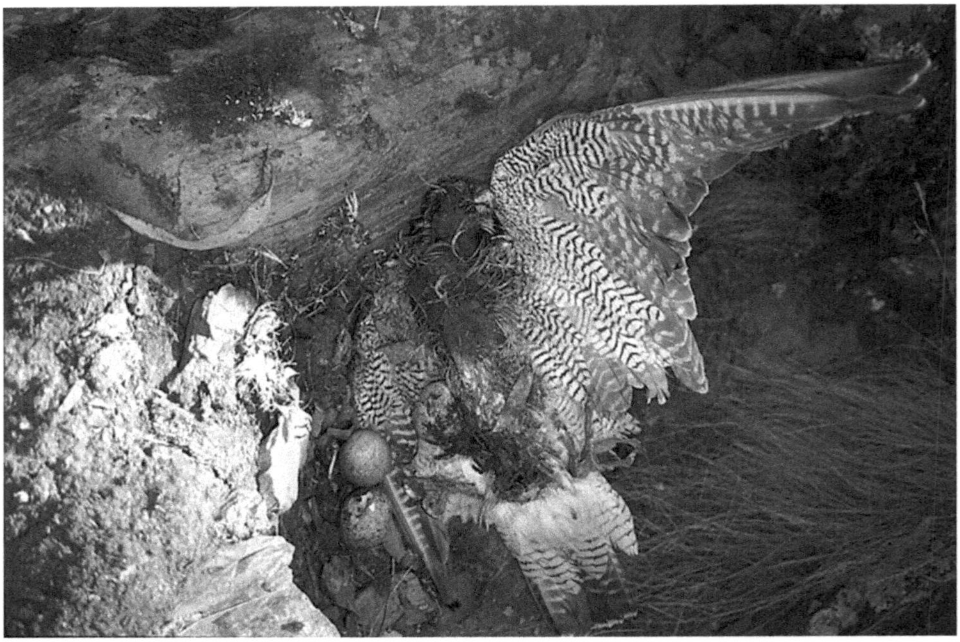

Figure 10.7 Female Peregrine dead on her eggs in NSW.

the lower 48 States, but Golden Eagles were not, resulting in at least one case of predator release where Golden Eagles moved into Bald Eagle-vacated islands off California and nearly wiped out the Island Grey Fox *Urocyon littoralis* there. Poisons can be species specific, and have system-wide consequences.

Prey availability versus selection

Raptors strike a balance between selecting prey that suits the raptor's size and hunting style versus prey that is most available. For example, Peregrines and Hobbies each select bird prey, but these are not necessarily the most abundant birds because Peregrines are larger than Hobbies and take larger birds, whereas Hobbies are smaller and more agile and take smaller more agile prey closer to cover. Neither falcon specialises on superabundant bird species such as Sulphur-crested Cockatoos because these are too heavy and dangerous to capture and kill. We monitored 16 Peregrine Falcon nest territories and one Australian Hobby territory in the Canberra region (Figure 10.8) during four breeding seasons in two time periods: 1991–1992 and 2002–2003, a total of 60 Peregrine nest-years and three Australian Hobby nest-years, 1991, 2002 and 2003 (Olsen *et al.* 2008). We looked at

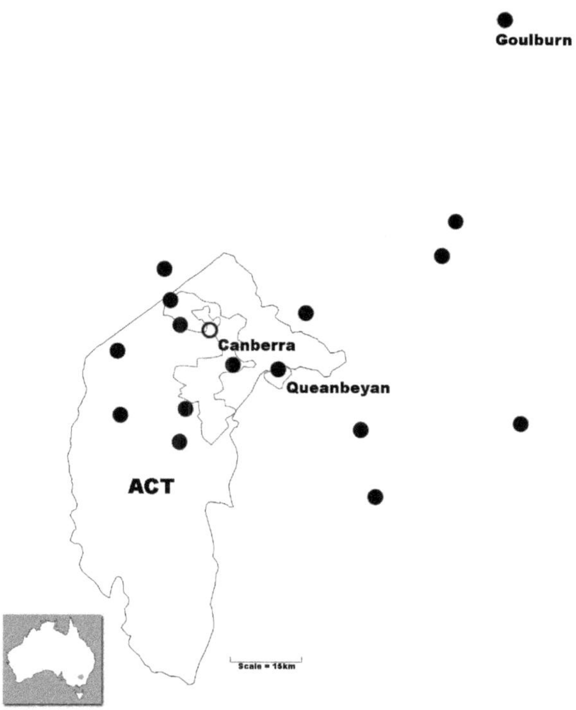

Figure 10.8 Locations of 16 Peregrine and one Australian Hobby territories monitored for prey during four breeding seasons – 1991, 1992, 2002 and 2003.

all the prey the falcons took, but particularly at their Main Prey Items (MPIs) defined as any species that contributed 5% or more by number to the overall diet in at least one of the two time periods (Table 10.1).

By number, Common Starlings were the main prey taken by both falcons in 1991–1992 and in 2002–2003, but starlings made up a smaller percentage of the diet by number in the 2002–2003 period than 11 years earlier, apparently because starling numbers had declined in the wild. We compared this capture rank to a Canberra Ornithologists Group's (COG) rank: the most common to least common prey species recorded by birdwatchers (Tables 10.1 and 10.2). Phillip Veerman (2003) related this decline in Common Starlings to loss of foraging habitat, and to a sharp increase in the population of Common Mynas, which fought with and displaced the starlings. Common Mynas were a perfect prey size and would have been available to both Peregrines and Hobbies at all nests in this study, but the falcons didn't take them, possibly because Mynas foraged too close to cover and were too agile. In contrast, Brown Goshawks, Southern Boobooks and Collared

Figure 10.9 Though Peregrines didn't take Common Mynas, Collared Sparrowhawks (pictured), Southern Boobooks and Brown Goshawks did.

Table 10.1 Main Prey Items (MPIs) by capture rank and weight (in grams) of species captured by Australian Hobbies and Peregrine Falcons

Capture rank	Australian Hobby	Species weight
1	Common Starling	75
2	House Sparrow	27
3	Crimson Rosella	135
4	Budgerigar	28
5	Eastern Rosella	106
6	Red-eyed Cicada	2
	Peregrine Falcon	
1	Common Starling	75
2	Galah	335
3	Eastern Rosella	106
4	Rock Dove	308
5	Silver Gull	303

Sparrowhawks did take Common Mynas in the ACT (Figure 10.9); Sparrowhawks used typical accipiter dashes low and fast to catch mynas a metre or so from protective cover.

Because Peregrines had larger sample sizes, we list the two time periods in Tables 10.2 and 10.3. The capture rank and COG rank are similar but don't quite match in either time period.

Another way to look at the same data, this time for both falcons, is by listing the ten most common bird species in COG surveys and looking at how Peregrines and Hobbies select from these (Table 10.4).

Table 10.2 Capture rank of birds taken by Peregrines and relative abundance in COG surveys 1991–1992

Capture rank		no. captured	COG rank	
		1991–1992	**1991–1992**	
1	Starling	164	*1*	Starling
2	Galah	70	*2*	House Sparrow
3	Eastern Rosella	48	*3*	Silvereye
4	Rock Dove	42	*4*	Galah
5	Silver Gull	34	*5*	Pied Currawong
6	Crimson Rosella	22	*6*	Sulphur-crested Cockatoo
7	Red-rumped Parrot	8	*7*	Common Myna
8	Magpie-lark	5	*8*	Crimson Rosella
9	Australian Magpie	5	*9*	Magpie
10	Australian Wood Duck	4	*10*	Yellow-faced Honeyeater

Table 10.3 Capture rank of birds taken by Peregrines and relative abundance in COG surveys 2002–2003

Capture rank		no. captured 2002–2003	COG rank 2002–2003	
1	Common Starling	121	*1*	Galah
2	Galah	101	*2*	Sulphur-crested Cockatoo
3	Rock Dove	66	*3*	Silvereye
4	Eastern Rosella	43	*4*	Common Myna
5	Crimson Rosella	22	*5*	Common Starling
6	Gang-gang Cockatoo	17	*6*	Pied Currawong
7	Red-rumped Parrot	12	*7*	House Sparrow
8	Red Wattlebird	10	*7*	Crimson Rosella
9	Magpie-lark	8	*9*	Magpie
10	Crested Pigeon	7	*10*	Wattlebird

Table 10.4 Ten most abundant species in COG surveys in 1991–1992 and 2002–2003 (from Veerman 2003; COG 2003, 2005), number of each species captured by Australian Hobbies and Peregrine Falcons in the two time periods, and the species capture rank (in brackets)

Ten most abundant species in COG survey, ranked in decreasing order of abundance for each time period		Australian Hobbies no. captured (rank)	Peregrine Falcons no. captured (rank)
		1991–1992	
1	Common Starling	31 (1)	164 (1)
2	House Sparrow	8 (2)	0
3	Silvereye	0	0
4	Galah	0	70 (2)
5	Pied Currawong	0	3 (10)
6	Sulphur-crested Cockatoo	0	0
7	Common Myna	0	0
8	Crimson Rosella	7 (3)	22 (6)
9	Australian Magpie	0	5 (8)
10	Yellow-faced Honeyeater	0	0
		2002–2003	
1	Galah	0	101 (2)
2	Sulphur-crested Cockatoo	0	2 (14)
3	Silvereye	1 (6)	3 (13)
4	Common Myna	0	0
5	Common Starling	13 (1)	121 (1)
6	Pied Currawong	0	1 (15)
7	House Sparrow	7 (2)	0
8	Crimson Rosella	0	22 (5)
9	Australian Magpie	0	4 (12)
10	Red Wattlebird	0	10 (8)

Peregrines regularly took some large species like Galahs, magpies and Silver Gulls, but Hobbies in this study took none of them. Common Starlings, Eastern Rosellas, and Red-rumped Parrots often forage in the open in flocks, and were specifically targeted by both falcons. Crimson Rosellas often forage in woodlands and were targeted less often by Peregrines but were targeted by Hobbies. Likewise, House Sparrows in the COG surveys were the second most abundant species in 1991–1992 and the seventh most abundant species in 2002–2003 (COG 2003, 2005). Hobbies in this study took them, but Peregrines did not. Budgerigars are seldom seen by ACT birdwatchers, but they were the fifth most common prey taken by Hobbies, obviously aviary escapees that didn't get far.

Both falcons compensated for the decline in Common Starlings in the latter period by taking a greater variety of bird species (Figures 10.10, 10.11), particularly small numbers of mostly native birds, rather than more of the one or two major prey species other than starlings. Peregrines took 37 bird species in the latter period that we didn't find in samples collected in the earlier period, and more individuals of some large species such as Gang-gangs, Galahs and Rock Doves. In the second time period, Peregrines took more than 100% more bird species and Hobbies took 50% more bird species. The decline in the number of Common Starlings in the falcons' diet and the increase in native species may be what is termed a 'functional response' to the decline of this exotic species, and relative

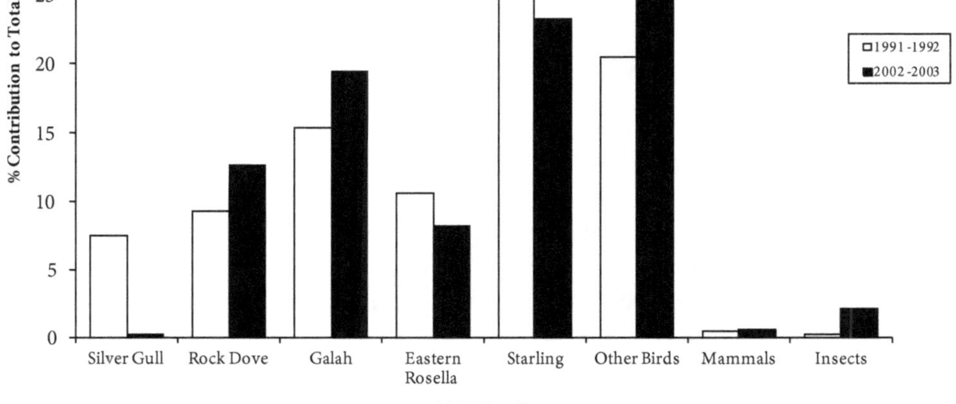

Figure 10.10 Comparative contribution of the Main Prey Items to the diet of the Peregrine Falcon in the Canberra region, 1991–92 and 2002–03.

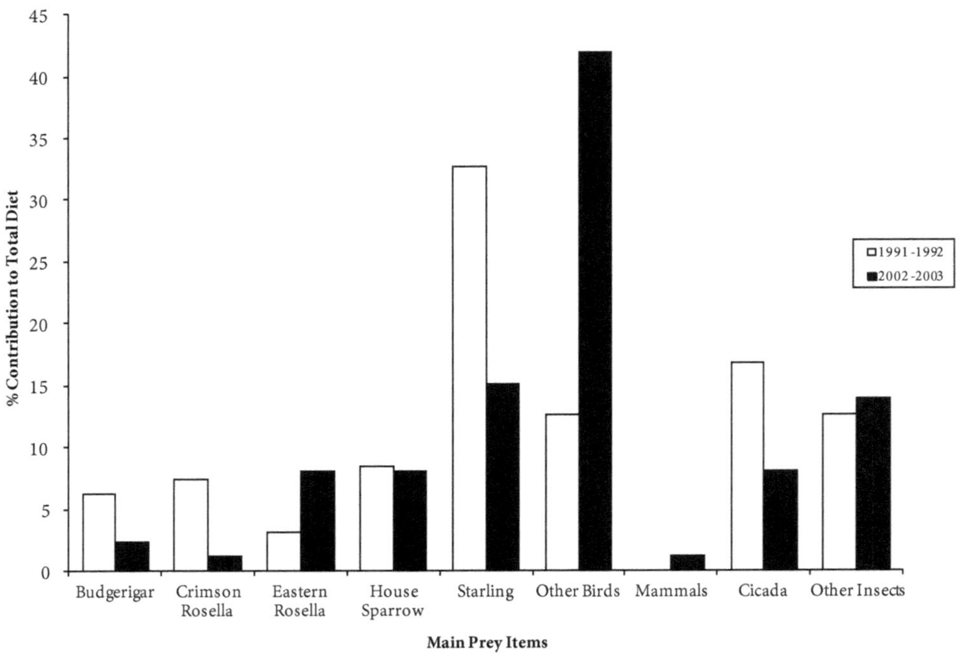

Figure 10.11 Comparative contribution of the Main Prey Items to the diet of the Australian Hobby in the Canberra region, 1991–92 and 2002–03.

abundance of similar-sized native species (Other Birds) in the area. For Peregrines, this may also be the case with increasing Galahs, Rock Doves, and Gang-gangs in the diet as Silver Gulls and Common Starlings declined.

The Geometric Mean Prey Weight of bird species taken by Hobbies was about one-third that of the weight of bird species taken by Peregrines, reflecting the difference in the two falcons' weight; Hobbies weigh ~33% as much as Peregrines.

Peregrine nest sites in the high country

Peregrine Falcons breed on all continents except Antarctica, and on many islands; they are perhaps the world's most widely distributed bird (Cade *et al.* 1988). As in all other *Falco*, they build no nest, but unlike many *Falco*, Peregrines throughout much of their worldwide range confine most of their breeding attempts to cliff ledges. Two limited resources, food and nest sites, largely determine the size and breeding success of a population of raptors (Newton 1979) so quality of nest sites is important. On average, ~57% of Peregrine sites around Canberra fledge young in a year, so ~43% fail or didn't try, and the average brood size (young per successful pair) is ~2.1 young each year. We could not link this to food availability, but we could link it to nest site availability and

Table 10.5 Average young fledged per territory from nests in potholes compared to nests on ledges

Nest type	Rainy year	Dry year
Ledge	0.91 young	1.43 young
Pothole	1.59 young	1.42 young

quality. Peregrines preferred 'weatherproof' sites that protected the adults, eggs and young from inclement weather; weather affecting the nest site is the best predictor of breeding success.

Peregrines in the high country fledged more young from a nest site protected from bad weather in a pothole or sheltered by an overhanging rock, or a nest ledge facing away from prevailing westerly storms. Peregrines around Canberra had plenty of food in the form of parrots, starlings and other birds but there was a shortage of nest sites, so some nested on substandard cliffs. The best territories had quality nest sites, and these fledged more young during rainy years, but there was little difference during dry years. We defined a rainy breeding season as August to October with more than 32 days with rain, a dry year as August to October with less than 32 days with rain. Peregrines that nested in potholes (Figure 10.12) fledged more young than those nesting on open ledges (Table 10.5), and those with

Figure 10.12 Peregrine nests in potholes fledged more young than Peregrine nests on open ledges.

Table 10.6 Mean number of young fledged per territory, % territories successful and % failed clutches for territories with more than one useable nest site compared to those with only one useable nest site

	One useable nest site		More than one useable nest site	
	Rainy year	Dry year	Rainy year	Dry year
Young per territory	0.66	1.25	1.68	1.45
% territories	35%	61%	75%	75%
% clutches failed	30%	18%	18%	6%

Table 10.7 Mean young fledged per territory from nest sites that do drain compared to nest sites that don't drain

Nest ledge	Rainy year	Dry year
Drains	1.04 young	1.39 young
Doesn't drain	0.72 young	1.43 young

alternative nest ledges to choose from in rainy years, about a third of the Peregrine territories around Canberra, fledged more young in rainy years than Peregrines that had only a single available nest ledge to use each year (Table 10.6).

The poorer nest sites did better in drought years than in rainy years because they were less likely to be flooded. One year, after the young were gone, I poured a litre of water into each of these sites to see if drainage mattered. It did. Sites that drained better, fledged more young (Table 10.7).

Near Canberra, where there are contiguous cliffs, Peregrines nested 2 km apart. This is as high as any inland density found anywhere in the world (Ratcliffe 1993). Because of the shortage of cliffs near Canberra, they will use lower cliffs than in most other parts of the world and we devised a sort of formula – a cliff near Canberra with the following characteristics usually had nesting Peregrines:

1 vertical;
2 at least 12 m high;
3 at least one suitable ledge;
4 at least 2 km away from other Peregrine nests;
5 no water flowing over the cliff;
6 not facing north-west; and
7 free from excess human disturbance.

This formula didn't hold at higher elevations, such as in Namadgi or Kosciuszko National Parks, where many cliffs fit this description but weren't used, apparently because the weather was too harsh.

Do Peregrines imprint on cliff or tree nests?

As mentioned above, most Peregrine nests found in Australia are on cliffs, but they also use stick nests in trees (Figure 10.13), and tree hollows. Overseas Peregrines seldom or never use tree hollows. Except for a few places along the Australian coast where White-bellied Sea-Eagles or Ospreys build nests on cliffs, Peregrines in Australia have no stick nests on cliffs to use, and no real competitors for cliff ledges inland. They are limited to tree hollows, stick nests in trees, or cliff ledges with a substrate of dirt, rocks or natural vegetation. How does an individual Peregrine choose to nest in a tree versus a cliff? Imprinting to the nest site is said to play a part. In Germany DDT wiped out most breeding Peregrines and when the falcons returned after DDT was banned, Germans built stick nests in baskets, placed these in trees and hacked (Olsen 1990) young Peregrines from these. However, Peregrines fledged from cliffs never moved to the baskets. This led to the conclusion by European biologists that imprinting fixed the nest site choice in Peregrines. The falcons nested on similar sites to their natal site, and disregarded potential tree nests that didn't physically resemble such a site (Kirmse 1994, 2004). 'Tradition founders' were defined as Peregrines that defy this principle and move from cliffs to trees or from trees to cliffs, and they only exhibit this exceptional behaviour in special and remote habitats, where the natal nest site is not available (Kirmse 2004). Kirmse argued that 'traditional founder' Peregrines, for example those nesting in tree hollows, are confined to a few isolated areas like southern Victoria in Australia, and the tradition is maintained by a lack of cliffs in the area, and by young fledged from tree sites that later choose to breed in tree sites instead of on cliff nests. However, there are only a few places in the world with remaining tree hollows big enough to serve as Peregrine nests. Victoria is an exception. For example, there used to be tree-nesting Peregrines in North America in places like Kansas and along the Mississippi River in Illinois and Indiana, but these have disappeared.

Peregrines in Victoria that nested in tree hollows and stick nests generally occurred in areas without cliffs, like the Mallee region, where they often use abandoned nests of Wedge-tailed Eagles. Less often they used tree nests of White-bellied Sea-Eagles, Little Eagles, Whistling Kites, cormorants and ravens (Marchant and Higgins 1993; Emison *et al.* 1997). Peregrines using stick nests in Victoria seemed to support Kirmse's argument because Wedge-tailed Eagles' nests are widespread and common throughout Victoria, but Peregrines don't use them, preferring cliffs where they are abundant.

Kirmse (2004) noted that recent colour-marking of Peregrines fledged in eastern Germany showed that tree nesting was never adopted by any offspring fledged from neighbouring buildings or cliffs. When Peregrines in Germany were imprinted to (fledged from) tree nests, they returned to breed in these tree nests, establishing six new breeding pairs in Germany in trees, the only known examples between Berlin and Moscow where thousands of pairs had existed before. Offspring from tree nests later bred in tree nests, but they also bred on buildings and cliffs. No Peregrines did the reverse (it is important to note that Kirmse considered a Peregrine fledged from a stick nest on a cliff as imprinted to cliffs). Kirmse (2004) concluded that:

Figure 10.13 Nestling Peregrine in an old Wedge-tailed Eagle nest at Lake George, NSW.

- In Peregrines tree nesting is based on imprinting and spreads by tradition.
- Offspring from cliffs or buildings will not adopt tree nesting.

In December 2003 a pair of Peregrines took over an abandoned Australian Raven's nest, previously used by Hobbies, 22 m up in a 27.8 m high Yellow Box *Eucalyptus melliodora* at Lanyon in the ACT (Figure 10.14). They raised young there that year and the next. The band on the male's leg identified it as a nestling I had banded on 29 Oct 1999 in a brood of two on a cliff ledge 23.4 km north-east of this stick nest. The brood fledged from a bare ledge 0.92 × 0.40 m, 15.5 m up a cliff annually used by Peregrines from at least 1987 to 2004 (Figure 10.15). It didn't look much like a raven's nest in a tree, but 5 years later this didn't seem to matter to the male. By 1 January 2005 the limb holding the nest blew down in a windstorm, and no Hobbies or Peregrines were seen. The Peregrine stick nest in 2004 was 9.2 km from the closest known Peregrine cliff nest and 57 km from the nearest known tree-nesting site used that year.

This was the first published report from anywhere in the world of a marked Peregrine fledging from a cliff and breeding in a tree nest, *contra* the predictions of Kirmse and the Germans. Why had this happened in the Australian high country but not in Europe? Peregrine numbers in Europe increased after pesticides were banned, but Peregrines still had cliffs available to them, especially where other

Figure 10.14 The adult male Peregrine on the left had fledged from a cliff ledge (see Figure 10.15), but nested in a tree, the first of its species shown to switch.

Figure 10.15 The adult male Peregrine in Figure 10.14 fledged from this bare ledge in Molonglo Gorge.

competing cliff-nesting raptors, like Eagle Owls, had been persecuted and reduced in number (Mikkola 1983). However, in Australia there was no measurable decline in the continental Peregrine population from pesticides, so cliff sites are saturated. As mentioned before, Australian Peregrines nest no higher than 1200 m above sea level, and the preferred cliff nest sites are limited. Tree nesting failed to naturally re-establish in Europe because Peregrines or other cliff-nesting competitors hadn't saturated the available cliffs.

Although both cliffs and trees are abundant in the high country, it is possible that strong competition for suitable cliffs below 1200 m pushed this male to choose a tree nest. In the high country where cliff nesters and tree nesters breed relatively close together, a tree-fledged Peregrine could mate with a cliff-fledged Peregrine. For example, near Lake George in 2004 one pair of Peregrines used a stick nest abandoned by Wedge-tailed Eagles and fledged one young; 1.9 km away another pair used a cliff nest but failed at the egg stage. These are, to our knowledge, the closest cliff and stick nests reported for Peregrines. One or another of a Peregrine pair breeding in a tree nest could lose its mate, and attract a partner fledged from a cliff that was unable to win a territory elsewhere, perhaps explaining why Peregrines in Australia, unlike Europe, readily mix cliff and stick nests.

11

Breeding and behaviour

Compared to many birds, the reproductive period of raptors is prolonged. Even the Southern Boobook, Australia's smallest owl, takes at least 5 months to complete a breeding effort, about the same length of time as a Peregrine Falcon. Kestrels and Black-shouldered Kites have quicker breeding cycles reflecting a more *r*-selected breeding strategy (i.e. more chicks, less care per chick) and taking advantage of irruptions of mammalian prey. Except for the Black-shouldered Kite, Canberra's raptors breed seasonally, once a year beginning in autumn (Powerful Owls), winter (large eagles), early spring (medium-sized falcons, small eagles) or late spring for other species. Most raptors establish nesting territories large enough to contain nests and roosts, and enough space for both the male and female to hunt in from incubation to post-fledging dispersal. Fierce territorial aggression early in the breeding season probably reflects this long-term goal. Brown Falcon males and females are particularly scrappy and noisy, kestrels too can openly battle with neighbours or other rivals. Wedge-tailed Eagles tend to aerial display, including 'pot-hook' displays, and so do Little Eagles all the while giving a characteristic whistling call, but both of these eagles can physically battle rivals (Figures 11.1, 11.2); Hobbies, Peregrines and accipiters all aerially display.

Raptors show fidelity to these defended territories, reoccupying them in consecutive years. It is common for one or both members of the pair to return annually to the same nesting area, often to the same nest site, until death or some environmental change causes disruption to this pattern (Craighead and Craighead 1956). Such fidelity to one breeding area enhances breeding success and the

Figure 11.1 Little Eagles battling over territory (Mark Osgood).

longevity of each raptor. They can learn all the features of a hunting area, where prey lives, feeds and takes refuge, and the location of obstacles, predators and other dangers that cause mortality. Raptors can also avoid overlapping territories with other raptors. One would expect the territories of insect specialists to be smaller than the territories of mammal or bird specialists due to factors such as density of prey, ease of capture and raptor body size (insect specialists tend to be smaller than raptors preying on mammals).

In species such as Wedge-tailed Eagles, Peregrine Falcons and Southern Boobooks, one or both members of the pair often remain year-round on the territory, though some Little Eagles seem to travel to a separate winter territory and some may migrate (Baker-Gabb and Fitzherbert 1989), and female Boobooks often desert fledglings leaving all parenting to males. Some species such as kestrels and Boobooks can breed when they are a year old. Some raptors, such as eagles or Powerful Owls, are said to form monogamous, lifelong bonds but this needs confirmation with long-term study. Pairs can remain bonded for many years, or until one dies, but studies of territories over many years, even decades, probably

Figure 11.2 Pre-breeding Wedge-tailed Eagles trying to drive off a caged pair of eagles inside their territory.

will show that some individuals switch territories, and mates. More likely, females change territories more often than males, and home-range sizes change over time (Olsen and Trost 2007).

During winter, male Southern Boobooks we studied remained on their breeding territories, but females often travelled 7 to 10 km away and spent the winter on a separate home range, or sometimes wintered in the far corner of their breeding territory. During winter, females often returned to the breeding territory and to last year's male for a night or two, and roosted with him. Returning females mutually preened with their mates and used low croaking calls in greeting, then returned to their winter home range. There is very little known about how Collared Sparrowhawks and Brown Goshawks winter but many probably migrate from the area. Our colleague Bill Mannan tells us that with closely related Cooper's Hawks in Arizona, males feed females year-round, and this could be the case with some Australian accipiters.

Courtship

At the beginning of the breeding season female raptors start food-begging, usually some form of whining or chittering, to induce males to provision them. These behaviours help drive the breeding system of raptors – males competing for females, and females choosing quality males and quality territories. Most female food-begging occurs before egg laying and during incubation and brooding, when males spend their days or nights hunting for the female and brood. If a male fails this initial food-provisioning test, or the female 'dislikes' something about the male, or about the territory, is stressed by the proximity of congeners, predators, or exposure to hostile weather, she moves on and tries again, or does not breed that year.

In the courtship period, some of this behaviour, including calling, allopreening and roosting together, occurs on borders or in different parts of the territory, but much of it occurs close to the nest. Courtship proceeds with long bouts of calling, often at the nest ledge, platform or hole, and with allopreening. The strong shows of 'affection' in owls (Figure 11.3) contrasts with courtship in diurnal raptors which are more standoffish with each other.

Owls face-preen their mates and sit with bodies touching, strange to anyone who has studied falcons or goshawks. In many diurnal raptors the sexes rarely touch each other and males often seem frightened of females. This underscores the phylogeny of different raptor groups – owls and goshawks may all kill vertebrate prey, but this is a convergent evolution in different basal groups of birds. In

Figure 11.3 Female Boobook face-preening her mate.

courtship behaviour, falcons fall somewhere between owls and goshawks, engaging in some 'affectionate' behaviour including billing. Square-tailed Kites and Black-breasted Buzzards do engage in allopreening. However, the claims that Wedge-tailed Eagles regularly engage in allopreening and courtship feeding have not come from published field studies, and need to be verified.

Nests

Owls and falcons build no nests. Instead they use tree hollows, rock cavities or the nests of other raptors and corvids. Other Australian raptors build nests distinctive in type and location to the species. Many, however, take over the nests of other species and repair these to suit. In the high country Whistling Kites and Little Eagles alternate use of certain nests, and Brown Goshawks take over raven nests and add to these nests for breeding. Raptors often build more than one nest and use these in alternate years. Wedge-tailed Eagles and Little Eagles use alternative nests or smaller nests as feeding platforms, or fledglings sometimes travel to these alternative nests soon after fledging, creating the false impression that young were raised there if the investigator first happens on this second nest with fledged young being fed. Black-shouldered Kites build a nest in spring, fledge young and, if prey is available, build a second nest and use it for a second brood. Some Wedge-tailed Eagles use only one nest and build onto it for many years until it is as big as a small van upturned. Sea-Eagles and Wedge-tailed Eagles will continue adding to a nest until the weight of the nest cracks the branch or tree and it plummets to the ground (Figure 11.4).

You often see raptors carrying sticks (Figure 11.5). Among Australian raptors, only the Swamp Harrier and Grass Owl habitually nest on the ground, though Peregrines can nest on rocky slopes. Nest-building skills vary between species and individuals; some young breeders construct scrappy, unstable platforms.

Finches and other small birds often nest in eagle and hawk nests and it has been argued that this is for the finches' protection. However, the hawk, eagle or falcon species that harbour such nests generally lack the agility to catch a finch or other small bird close to the nest, and the finch knows the raptor's daily habits and can avoid being caught. You will seldom find finches nesting in a goshawk or sparrowhawk nest because these accipiters could easily snatch up a small bird close to or in the hawk's nest. One way to test whether this habit is really for the finch's protection or not is to observe how often it occurs in used versus abandoned hawk, eagle, falcon and corvid nests.

Incubation

Owl eggs are white, falcon eggs are blotched with rusty colours and browns, eagle and hawk eggs are white variably blotched with browns and ochres. Compared to

Figure 11.4 This Wedge-tailed Eagle nest was added to each year until it eventually broke the branch.

other raptors' eggs which are shaped more or less like a chicken egg, owls' eggs are comparatively round, a shape that occupies less space for their volume than do other configurations, as one would expect for hollow-nesting species.

Raptors lay eggs at 2 to 3 day intervals, depending on the species. Some raptors such as harriers and *Tyto* owls begin incubation from the first egg so the nest can be filled with different-aged young, but many other raptors begin incubation with the penultimate egg. This means that nestlings in a brood of falcons, accipiters or eagles will have one noticeably younger chick, but the other chicks are closer to the same age. Most eagles lay only one or two eggs. In the days before egg laying, females sit fluffed-up and looking ill. This is called 'egg-laying lethargy'. This period of inactivity allows eggs to form and gives the appearance of lethargy in the female, but this behaviour stops after eggs are laid.

Generally speaking, large species such as eagles have smaller clutches, often two eggs, while smaller species such as kestrels and sparrowhawks have larger clutches, up to seven eggs. However, the clutch size in high country raptors varies

Figure 11.5 Wedge-tailed Eagle carrying sticks.

with each species, and the difference is not closely linked with the body mass of the raptor (Table 11.1). The 14 species that breed in the high country, plus the Barn Owl which we may eventually find nesting here, show little variation in clutch size, though Barn Owls, Black-shouldered Kites, Brown Falcons and kestrels show a greater range in clutch size than do Peregrines, accipiters, eagles or Powerful Owls. These latter species also tend to have smaller clutch sizes than do rodent specialists such as Black-shouldered Kites and Barn Owls. Those with larger clutch sizes and ranges are said to be typical of *stenophagous* raptors, that is, raptors restricted to a small number of prey species. This contrasts with *euryphagous* species, raptors that have a broad prey base, such as Southern Boobooks, Brown Falcons, Peregrines and eagles, which have smaller clutch sizes and ranges, but the contrast is not as marked as in Northern Hemisphere raptors. Species such as Brown Falcons seem to have relatively consistent clutch sizes in Victoria that remain small in good and poor years (McDonald *et al.* 2003). However, there may be other factors at play besides width of prey base. Kavanagh (2002a) said that Powerful Owls took 62%

Ringtail Possums by number, and Debus *et al.* (2004) said that Barn Owls took 62% House Mouse by number, but these two owl species have significant differences in clutch size and ranges, and it may be related to their taxonomic origins (*Ninox* and *Tyto*). Probably *stenophagous* owls rely on a small number of highly fluctuating key species, whereas *euryphagous* species do not. Populations of key species such as the House Mouse or Long-haired Rat *Rattus villosissimus* fluctuate much more than those of Ringtail Possums. The key factor may not be the number of key prey species that a hawk, falcon, eagle or owl takes, but how much the key prey species' populations fluctuate or cycle.

Powerful Owls, Wedge-tailed Eagles, Little Eagles and Peregrines may be more *K-selected* than are species such as Black-shouldered Kites and Barn Owls that are more *r-selected*. K-selected species are long-lived, slow reproducing and their populations stay fairly constant near the carrying capacity (K) of relatively stable environments. One might expect this in nesting south-eastern highland raptors where conditions fluctuate (relatively) little across years. In contrast, Barn Owls, Black-shouldered Kites, kestrels and maybe Brown Falcons, especially in the arid inland, can be more *r-selected*, have potentially broader clutch ranges and brood sizes, reproductive rates and dispersal systems that allow for rapid changes in population sizes, and the ability to change locations. Barn Owls and Black-shouldered Kites can adapt to more unpredictable environments and carrying capacities, such as Australian grasslands and the arid inland.

Some male raptors incubate, especially falcons, but male owls never incubate, at least in the species so far studied. In diurnal raptors, such as falcons or eagles, males incubate as much as females allow them to. During incubation the male normally feeds the female by hunting some distance from the nest, or he guards territory borders, and she defends the nest against predators. However, males do defend nests when they are present. Male Peregrines will sometimes defend against humans then appear to be drawn away from the nest to hunt, leaving nest defence to the female. Incubating female Southern Boobooks hunt right through the incubation period, sometimes leaving their eggs for up to an hour when outside temperatures are close to 5°C. Apparently, eggs are protected by the insulated tree hollow, and wood chips further insulate eggs and nestlings. Female Boobooks hunt insects close to the nest while males search farther afield for less common vertebrates. However, falcon eggs exposed on a cliff ledge, or hawk or eagle eggs left exposed in an open nest, would soon chill or be taken by predators.

Raptors are very careful with their lethal talons when incubating and brooding. They fold the talons in and stretch the toes forward when resting on eggs, and they regularly turn the eggs carefully with their beaks.

Incubating diurnal raptors are extremely secretive; they lie flat on the nest if people approach (Figures 11.6, 11.7). Sometimes it can be difficult to determine if the nest is occupied. Raptors are extremely sensitive to disturbance in the early

Table 11.1 Clutch size, main food and habitat types of high country raptors

Species	Female mass (g)	Clutch size		Main food types	Main habitat type
		mean	range		
Black-shouldered Kite	300	3.4	2–5	ground mammals	grassland, open woodland
Whistling Kite	850	2.0	1–4	mammals, fish, birds, invertebrates	open woodlands, grasslands, riparian
White-bellied Sea-Eagle	3800	2.0	1–3	riparian/marine birds, reptiles, fish	rivers, dams, sea shores
Swamp Harrier	850	3.6	2–7	small mammals, carrion, birds, eggs, insects, frogs, fish, reptiles	terrestrial wetlands, fresh or salt
Grey Goshawk	720	2.4	2–4	birds, mammals, reptiles, invertebrates	dense forest
Brown Goshawk	570	3.0	2–4	birds, mammals, reptiles, invertebrates	woodland, forest
Collared Sparrowhawk	240	3.2	2–5	birds, invertebrates	woodland, forest
Wedge-tailed Eagle	3800	2.0	1–3	medium and large mammals, birds	open grasslands and woodlands
Little Eagle	1113	1.7	1–3	medium-sized mammals and birds	open grasslands and woodlands
Brown Falcon	625	2.7	1–5	medium-sized mammals, birds, insects, snakes, lizards, rabbits, carrion	open grasslands and woodlands
Nankeen Kestrel	185	4.3	1–6	terrestrial vertebrates and invertebrates, birds, lizards, mice	open grasslands and woodlands
Australian Hobby	290	2.8	2–4	birds, insects	open grasslands, woodlands, mountains
Peregrine Falcon	890	2.9	1–5	birds, insects	open riparian, grasslands, woodlands, mountains
Barn Owl	360	4.0	1–7	ground mammals	open woodland, grassland
Masked Owl	835	1.6	1–4	ground mammals	forest and woodland
Sooty Owl	1250	1.5	1–2	arboreal and ground mammals	dense forest
Powerful Owl	1350	1.9	1–2	arboreal mammals, bats, birds	forest
Southern Boobook	315	2.5	1–4	birds, insects, ground mammals	forest and woodland

Figure 11.6 Incubating Brown Goshawk watching the author.

Figure 11.7 Incubating Wedge-tailed Eagles lie flat when humans approach, but flush if disturbed.

stages of incubation and may desert the nest; this is especially the case with Swamp Harriers. Females of all raptor species sit especially tight when the eggs are close to hatching. Even large *Aquila* eagles, the wariest of birds of prey, will sit tight during this time with humans nearby, and many female Collared Sparrowhawks or Southern Boobooks refuse to leave hatching eggs at all. They can be lifted off the clutch, banded, and will resume incubating or brooding when the person applying the band releases them. However, we do not advise this level of disturbance.

Nestlings

Newton (1979) described three overlapping phases of female behaviour typical for breeding raptors with nestlings (Figure 11.8): 1) almost continuous brooding or shading before the young have developed their own thermoregulation; 2) presence near the nest for feeding and defending the largely helpless young (Figure 11.9); and 3) attendance with flexible periods of time spent distant from older nestlings or fledged broods. Attendance patterns during the third phase varied greatly, and Newton suggested that the variation depends largely upon the hunting success of the male, but it can also depend on the female's perceived threats to the nestlings from enemies and weather.

Figure 11.8 Newly hatched Peregrines.

Figure 11.9 Female Collared Sparrowhawk feeding small young.

Falcons especially defend their nestlings against predators such as humans, but different falcon species use slightly different attack approaches (Figures 11.10, 11.11).

Females seem to leave broods unprotected only when the males' hunting efforts are insufficient and chicks require additional nourishment. However, Southern Boobooks, and perhaps all *Ninox* owls, break these rules. Female Boobooks hunt through incubation and the early nestling phase.

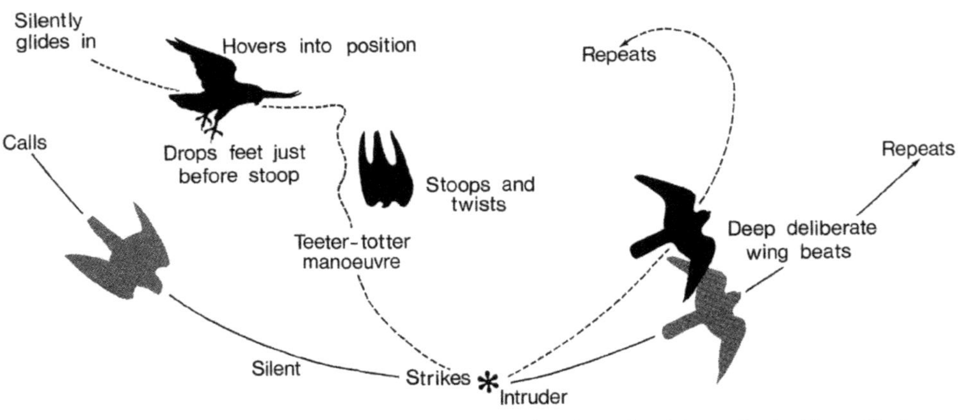

Figure 11.10 Pattern of defence against human intruders in four Australian falcons. Dashed line = Nankeen Kestrel and Black Falcon. Solid line = Peregrine Falcon and Australian Hobby. Hovering into position only occurs in the Nankeen Kestrel. The 'teeter-totter manoeuvre' (Balgooyen 1976) is a twisting of less than 180° about the long axis of the body. After Olsen and Olsen (1980a).

Figure 11.11 Nankeen Kestrel defending its nest.

Female raptors need to guard the nest and young, but some female Boobooks could guard and hunt at the same time, if prey was close by.

Males would sometimes arrive with prey, call to the female, go into the hollow, fail to find her, and then go looking for the female instead of giving the prey item to his nestlings. Researchers interpreted this as a male 'showing' his mate that he was providing large prey for his time away, even if it was a large-looking moth. This might encourage her to stay during that breeding season and return the following year.

Except for the eagles and large *Ninox* and *Tyto* owls, raptor nestlings grow quickly. Most start with a dense white down that is replaced by a sparser natal down that is soon penetrated by pinfeathers on the wings and tail. These juvenile feathers grow more quickly than body feathers, giving many nestlings a mottled or two-tone look. At this point many young require less brooding to keep them warm,

Figure 11.12 Nestling Brown Goshawks at this age are curious; they watch moving animals and play with objects in the nest.

less shading from the sun and less protection from predators, so females begin hunting further from the nest, leaving the young on their own for longer and longer periods. At this point larger prey starts showing up at the nest, for example, more Galahs at Peregrine nests, though many male and female raptors take prey that is about the same size. Nestlings are powerful enough to pull apart large prey and feed themselves, but, if parents are present, nestlings and fledglings act like babies and call to be fed. Nestlings track anything that moves (Figure 11.12) – birds, butterflies, beetles – and they snatch at twigs and prey items in the nest, practising killing. Nestling eagles at this stage have a pleasant sweet smell and nice breath, unlike mammalian carnivores.

In raptors with Reversed Sexual Dimorphism the smaller males develop faster than females. As they approach fledging age, young hawks, eagles and falcons flap and exercise their wings, and jump up and down on the nest; some are occasionally carried off by stray winds and end up on the ground. Nestling Southern Boobooks often crawl outside and face into the nest hollow and flap their wings like little hurricanes. There is only room for one nestling at a time to do this, so one will tire and climb back into the nest, and a second will climb out and take its turn whirring its wings.

Except for feather growth, young raptors are essentially fully grown at the fledging stage, and they can defend themselves from enemies, including humans, by flipping onto their backs and presenting taloned feet. These talons can do a lot of damage.

First flights are usually clumsy with poor landings. Even if the young land a little way from the nest most parents feed them, not because they recognise them (though parent Boobooks do recognise their own young), but because parents are driven to feed any young of the same species close to the nest.

Because diurnal raptors don't recognise their own young (though usually they can discriminate their own species), researchers have placed captive-bred Peregrines and Wedge-tailed Eagles into the nests of wild Peregrines (Figure 11.13)

Figure 11.13 Captive-bred or rehabilitated nestlings of slightly different ages from those in a wild nest can be successfully fostered.

Table 11.2 Estimating nestling age from wing length (after Olsen and Olsen 1987; Ridpath and Brooker 1986; McDonald 2003b; and J Olsen and S Trost unpublished data)

Species	Ageing formula
Black-shouldered Kite	(Wing length + 0.90) /0.65
Whistling Kite	(Wing length + 1.47) /0.59
Brown Goshawk	(Wing length – 1.12) /6.25
Collared Sparrowhawk	(Wing length + 0.42) /0.59
Little Eagle	(Wing length + 0.42) /0.57
Wedge-tailed Eagle	(Wing length – 2.37) /0.60
Peregrine Falcon	(Wing length + 0.84) /0.69
Australian Hobby	(Wing length + 0.42) /0.59
Brown Falcon	(Wing length + 19.4) /6.81
Nankeen Kestrel	(Wing length + 0.76) /0.62
Southern Boobook	(Wing length + 1.52) /0.52

and Wedge-tailed Eagles, and these were raised as normal. The point is that parents will feed displaced young around nest sites, and some researchers even nail artificial nests to the tree-trunk below the raptor's actual nest so they, the researchers, can tether nestlings there and readily access and measure the young without climbing a high tree. Parents usually feed these nestlings.

In Table 11.2 above are formulae for estimating the approximate nestling age from wing length for some high country raptors, best used after the pinfeathers have emerged. For calculating nestling age, place a ruler under the unflattened wing and measure from the bend of the wing (Figure 11.14) to the end of the wingtip, then apply the formula to find the nestling's approximate age. For example, a Boobook nestling with a wing length of 10 cm would be calculated as 10 + 5.2 = 15.2 divided by 0.52 gives an age of 29 days.

Fledging

There is some confusion in the literature around the definition of fledging. According to one standard text, *A Dictionary of Birds*, fledging means a bird leaving the nest when it is ready to fly and has attained its first complement of true feathers (Campbell and Lack 1985). However, owls and some other birds can leave the nest before they fly, and before flight feathers and tail feathers are fully grown. If these birds have left the nest but are not fledged, what is the name of this stage? One answer is 'branching'.

Owls, and some hawks, especially accipiters, and eagles, have young that 'branch' into foliage adjoining the nest. Owlets and goshawks crawl out along branches flapping their wings to balance and using their claws and sometimes their beaks to move out to the ends of limbs, then back to the nest. Owls sometimes

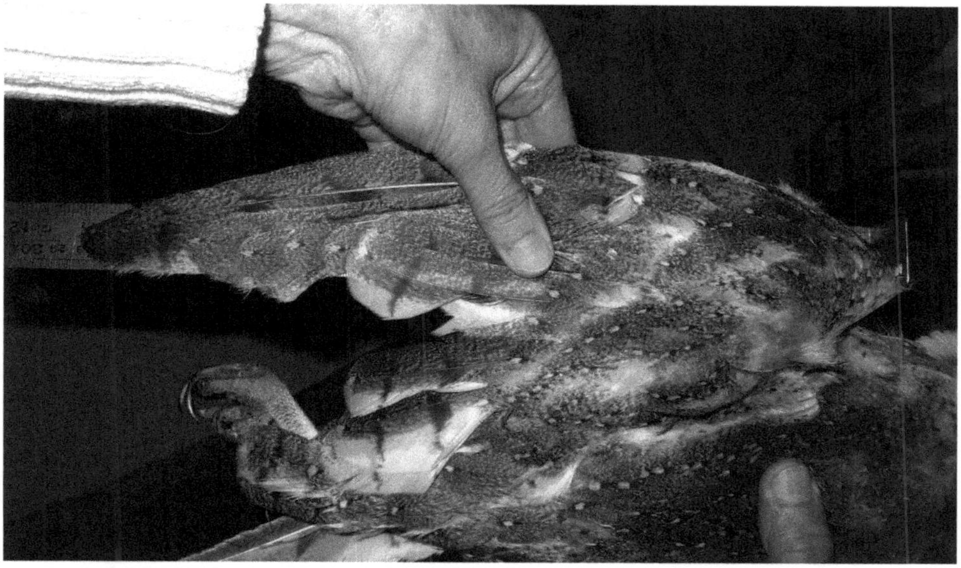

Figure 11.14 Measuring wing length of a Barn Owl.

slip at this stage and fall to the ground, or land further down in an adjoining tree. Humans can pick them up and they end up in rehabilitation centres, where they should not be. Owlets can glide, but they cannot fly. They often try to get up off the ground by walking to a nearby tree and climbing up, using beak, talons and flapping wings to get as high as they can. Finders should pick up these young owls, or any hawks, eagles or falcons they find, and place them in a tree close to where they were found; the parents will find them. With owls, do this when it is dark so the young ones are not attacked by enemy birds.

Ian Newton (1986) differentiated 'branching' from 'fledging' in Eurasian Sparrowhawks. Nestling males move onto branches near the nest at 19 days old, the larger females at 21 days old. Males make their first flight at 26 days, females at 30 days. Collared Sparrowhawks are about the same.

Olsen (2011a) found that Boobooks fledged later if the nest hollow was in a smooth-barked tree like a Scribbly Gum (difficult for owlets to climb up), and if there were no branches or ledges nearby for the owls to crawl to, where they can walk, receive food, then move back into the nest for the day. Older nestlings with no ledge or close branch sometimes fixate on a branch a metre or two from the nest, study it for several minutes while perched in the nest entrance, then suddenly fly to the branch and land clumsily. Young Southern Boobooks from around 4.5 weeks climb out of the nest hollow at night and back into the nest for the day, before official fledging closer to 5.5 to 6 weeks. Black-shouldered Kites branch in the last 4 to 7 days of the nestling period; Whistling Kites at ~32 to 46 days of age; in the second half of the nestling period Swamp Harrier nestlings clamber out of

Table 11.3 Ages when nestlings fledge for raptor species that breed in the high country cited in Marchant and Higgins (1993), Higgins (1999) and Debus (2012)

Species	Fledging age
Black-shouldered Kite	~35 days (33 to 38 days)
Whistling Kite	~44 to 54 days
White-bellied Sea-Eagle	~81 to 84 days
Swamp Harrier	~43 to 46 days
Grey Goshawk	~30 to 38 days
Brown Goshawk	males ~30 days (28 to 31 days), females ~31 days (28 to 37 days)
Collared Sparrowhawk	~28 to 33 days
Wedge-tailed Eagle	~80 days (77 to 95 days)
Little Eagle	~60 days (52 to 66 days)
Brown Falcon	~37 days for males, 39 days for females (36 to 42 days)
Nankeen Kestrel	~31 to 35 days
Australian Hobby	~36 days (34 to 41 days)
Peregrine Falcon	~42 days (38 to 45 days)
Eastern Barn Owl	~45 to 55 days
Masked Owl	~10 to 12 weeks
Sooty Owl	~3 months
Powerful Owl	~7 to 9 weeks
Southern Boobook	~6 weeks (28 to 45 days)

the nest and seek shade; Brown Goshawks stand on nearby limbs after ~20 days; Collared Sparrowhawks clamber onto limbs at ~26 days and maybe earlier; Wedge-tailed Eagles ~72 days, but there is great variation; Little Eagles tend not to branch as early as other species, often staying in the nest until they can fly; Brown Falcons tend to stay nest-bound but some branch around 35 days; Nankeen Kestrels tend to peek out from the nest hollow in the weeks before fledging, though some in cliff nests wander along cliff ledges; Peregrine young wander onto ledges and rocks on the cliffs, often after ~3.5 weeks.

The ages when nestlings fledge for raptor species that breed in the high country cited by Marchant and Higgins (1993) and Higgins (1999) (Table 11.3) should be treated with caution, particularly because of different definitions of fledging, and because most fledged young of high country raptors return to the nest for feeding so observers see them and assume they have not already fledged.

Fledgling behaviour

Fledgling behaviour is poorly understood in raptors. The period between a juvenile's first flight from the nest and its dispersal from the breeding territory has received little attention. It is a crucial period for young raptors because they need to practise hunting at this time. If juveniles fail to acquire hunting skills during this period they may starve, even in the face of abundant food. Most do manage to attain them, but they need to have done so before winter cold and food shortage make life hard, at least in the high country.

The post-fledging period of most raptors is structured around the nest site – adults return to the nest with food, deliver it, sometimes feeding the young, or simply dropping it and leaving. Fledglings 'know' that prey will be delivered to the nest site so they stay close and watch for it. Boobooks and some other *Ninox* owls are different. They travel through the woods and camp out at various locations, sometimes close to the nest, sometimes far from the nest, and keep track of each other through persistent calling. Peregrines and Wedge-tailed Eagle families sometimes move to a location away from the nest after young fledge.

Parents of some species, such as Peregrines, bring live prey and drop it for young to practise on and catch, but there are two parallel learning systems operating here. Hacked young (released systematically from a hack box) learn to hunt prey without help from their parents; they learn through curiosity and experimentation, much as a domestic cat's kitten learns how to hunt. We never saw Southern Boobooks bring live prey for their young to practise on, though fledglings grabbed live beetles and other invertebrates.

Values for fledging to independence given for high country raptors taken from Higgins (1999) are given in Table 11.4.

The most obvious behaviour in fledged raptors is food-begging. Most species use whining, churring, rasping or other calls. Boobooks use a cricket-like 'Trill' and they do this persistently through the night. The 'Trill' call is also a contact call, allowing young owls to find each other at night and stay together. During the day owls may roost back inside the nest hollow, in a different hollow, or with all or part of the family in a eucalypt, exotic tree or Native Cherry. In some species, such as Boobooks, the female deserts the young midway during the fledgling stage, leaving all parental duties to the male. Fledged Peregrines often return to the nest or another ledge on the cliff, and fledged eagles often start roosting and receiving prey at another roost tree away from the nest but in sight of it.

Dispersing young raptors will sometimes roost close to another brood of fledged juveniles for a time. Sue Trost and I saw a pale-coloured juvenile Australian Hobby, apparently from inland Australia, join a brood of three recently fledged darker young near Canberra. Young falcons, owls and other raptors appear to seek out the company of others of the same species. Adults and juveniles usually tolerate them. Adult owls usually don't feed these 'strays' but neither do they drive them off. In 2012 we introduced a flying but inexperienced juvenile Southern Boobook to an unrelated family of just-fledged Boobooks. We never saw an adult feed this owl but the rehab juvenile seemed to learn to hunt aerial and terrestrial prey by

Table 11.4 Values for fledging to independence in high country raptors cited by Marchant and Higgins (1993), Higgins (1999) and Debus (2012)

Species	Time from fledging to independence
Black-shouldered Kite	~3 or 4 weeks
Whistling Kite	~6 to 8 weeks
White-bellied Sea-Eagle	~2 to 3 months
Swamp Harrier	~4 to 6 weeks
Grey Goshawk	up to 6 weeks
Brown Goshawk	up to 6 weeks
Collared Sparrowhawk	~ up to 6 weeks
Wedge-tailed Eagle	~3 to 6 months
Little Eagle	~8 weeks
Brown Falcon	up to 6 weeks
Nankeen Kestrel	up to 8 weeks
Australian Hobby	~11 to 12 weeks
Peregrine Falcon	~11 to 12 weeks
Eastern Barn Owl	~12 weeks
Masked Owl	~10 to 12 weeks
Sooty Owl	~3 to 5 months
Southern Boobook	6 to 7 weeks
Powerful Owl	150 to 240 days

watching its step-siblings, following them to the ground, capturing moths after watching them catch moths, and so on. Ed McNabb (1994) successfully fostered a fledgling Powerful Owl in Victoria with a pair of owls that had a single similar-aged young but did not see the adults feed it.

Parent raptors eventually withhold food from their fledged young and the young leave, though again this is a parallel system because hacked young disperse as normal even though humans supply food each day. It must be related to some internal clock, perhaps influenced by weather, and competence at finding food, i.e. invertebrates, vertebrates or carrion in the case of large eagles and kites.

Timing of breeding

Many raptors in south-eastern Australia breed in late winter/early spring, and fledge their young in late spring/early summer, commensurate with a flush of available prey. Ornithologists find it useful to separate 'proximate' factors, the actual *triggers* that cause raptors to start breeding from 'ultimate' factors, the *reason* why they breed when they do. In south-eastern Australia day-length is the most important external cue that kicks off breeding in late winter for most raptors. Prey abundance for most species in Canberra goes through a bottleneck – lowest in winter when spiders, insects and birds are least available, and peaking in late summer when beetles, moths, spiders, birds, rabbits and the fledged young of parrots and small passerines are most available. At the end of summer, food supply declines again and becomes scarce, hence the winter bottleneck. This is a potential problem because raptors need enough food to bring them into breeding condition. Close to the winter, Canberra raptors start pre-breeding behaviour so they can fledge their young by the summer peak in food supply. One prey exception could be the House Mouse, which can peak in winter, and this explains why the mouse-specialist Black-shouldered Kite can breed in spring, then start again in autumn, fledging young in the teeth of the high country winter. Powerful Owls also start breeding in autumn, apparently because of the breeding times of their arboreal marsupial prey.

With most birds of prey, changes in day-length stimulate gonadal growth and increase female appetite. She responds by catching more prey and accepting extra food brought by the male. She lays down fat reserves and will experience growth in both the ovary and oviduct, and begin yolk deposition. Males will experience growth of the testes, but, in most species, only female raptors lose feathers from their lower abdomen and form a brood patch, an area of vascularised skin through which heat is transmitted to eggs and nestlings during incubation and brooding. Falcons have two brood patches. This is probably why there is a 'false start' in breeding of many birds, including raptors such as kestrels, during the high country autumn, with calling, agonistic behaviour, even courting – this day-length mimics the day-length of early spring. In the calling behaviour of Southern Boobooks (Figures 11.15, 11.16) there is also a smaller spike in April mirroring the breeding activity in diurnal raptors.

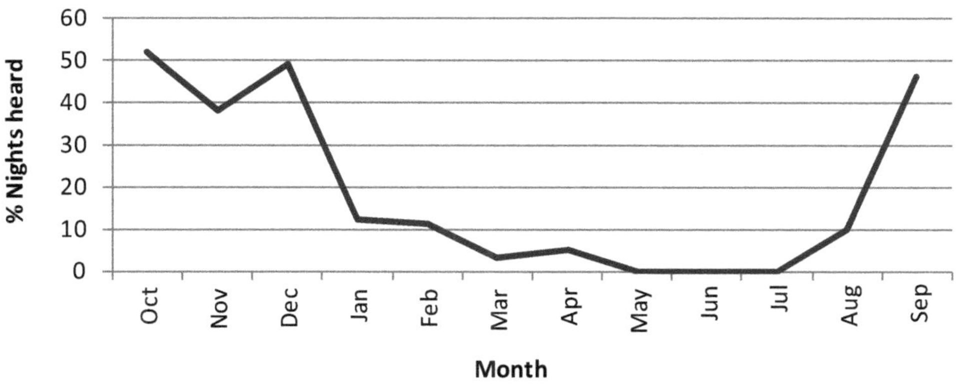

Figure 11.15 Percentage of nights when territorial singing by Southern Boobooks was heard in the ACT.

Figure 11.16 Owls such as Southern Boobooks defend territories by calling.

Ultimate factors, that is, the ultimate reasons for having eggs, nestlings or fledglings during spring or summer in the high country, can include food supply in a particular season. It is a convention in the birds-of-prey literature to assume that the ultimate factor in the timing of raptor breeding, the reason why they breed when they do, is so peak food supply occurs during the nestling period (Newton 1979). However, Olsen and Georges (1993) argued that the ultimate timing for breeding Peregrines might relate to the period when young raptors have left the nest and are learning to hunt. When most young raptors fledge, they cannot successfully capture enough prey to sustain themselves, so parents continue to feed them. Before their parents stop feeding them, fledglings need practice in capturing birds, mammals or insects that are their main prey; otherwise they will starve. For example, in Figure 11.17, four main prey species of Peregrines – Common Starlings, Galahs, Crimson Rosellas and Eastern Rosellas – have a peak in abundance, because their own young have fledged, around the same time that Peregrine young fledge, not when Peregrines have nestlings.

```
                        Aug  Sept  Oct  Nov  Dec  Jan  Feb  Mar

Starling                                 I---------------I
Galah                                         I---------------------I
Crimson Rosella                               I---------------I
Eastern Rosella                                    I----------I

Peregrine                          I-------I  I-----I
                                   hatch      fledge
```

Figure 11.17 Fledging of Peregrines and prey species in Canberra 1984.

In summary, the main proximate external trigger for breeding is day-length. But because of ultimate factors, the length of the breeding cycle is important in timing its start. If the breeding cycle of Powerful Owls or Wedge-tailed Eagles (Figure 11.18) is 10 or 11 months long, they no sooner get rid of dependent young before they start the cycle over again, because their juveniles remain dependent for a long time after they fledge. A key to understanding the timing of breeding in many raptors may be the length of this fledgling stage. The longer the fledgling period, the earlier the raptor will breed. Raptors breed early enough in the year to complete a breeding cycle before winter food shortages kill individual raptors and reduce prey, and this influences the time breeding is initiated.

Figure 11.18 Molonglo River Wedge-tailed Eagle nest with single young just branching.

12
Why are female raptors bigger than males?

In most birds and mammals male–male competition has led to males being larger than females ('Normal' Sexual Dimorphism, NSD). However, in birds of prey the opposite is true; most exhibit Reversed Sexual Dimorphism (RSD) with females larger than males. Over 20 hypotheses have been proposed and debated in the literature to explain RSD in raptors (Krüger 2005). In his analysis, Krüger said these hypotheses fit into three broad theories: *niche partitioning* – sexes diverge in size to reduce intersexual competition for prey; *role differentiation* – females become larger to protect and efficiently incubate eggs, and/or males become smaller for more efficient foraging or territory defence; and *behavioural* – larger females dominate males (Figure 12.1), aid in the maintenance of the pair bond and increase male food provisioning *or* larger females compete more effectively with other females to win males. Krüger collected data on 237 Accipitridae (hawks), 61 Falconidae (falcons) and 212 Strigidae (owls), and determined a dimorphism index from the wing length of males (mm) divided by wing length of females. He performed comparative analyses, using both cross-taxa data and phylogenetically independent contrasts, to investigate potential correlates of RSD. Using a set of explanatory variables, covering morphology, life history and ecology, he analysed 26 predictor values for hawks and falcons and 22 for owls, with RSD as the

Figure 12.1 Female Wedge-tailed Eagles are larger than males and may control food from them (Mark Osgood).

dependent variable. He found no evidence linking RSD to egg formation, egg size or females competing for males (sexual selection). He did not discuss an additional problem with the female–female competition hypothesis – males are often observed competing for females, for example in hooting owls (Olsen 2011a) but there is no spike in female–female conflict during the pre-breeding or breeding season, as often seen with territorial males or territorial pairs. For example, when easily observed migratory species like Black Kites arrive in Spain from Africa, there is no spike in visible female–female fighting (Fabrizio Sergio, pers. comm.).

Instead, Krüger found evidence that RSD might have developed for hunting large and agile prey, providing support for the *intersexual-competition hypotheses* – sexes have evolved different sizes to reduce competition for prey, and the *small male hypotheses* – males became smaller to be more efficient foragers. Krüger found in his analysis of owls that Australia held the two extremes: Sooty Owl exhibits the most RSD and Rufous Owls the least (because the Rufous Owl, as with the Powerful Owl exhibits NSD, with males larger than females). He did not discuss this anomaly.

Costs and benefits of hooked beaks and curved talons

Raptors differ from other birds in having hooked beaks and deeply curved talons. A factor never considered in previous studies of RSD is the *cost* and *benefits* of hooked beaks and curved talons. By contrast, a factor that is commonly emphasised in studies of RSD is the importance of 'production feeding' of female raptors by males in the early breeding cycle (Newton 1986; Balgooyen 1976) and the male's role in feeding the female and nestlings early in the nestling period. However, male corvids such as the Common Raven also feed females before egg laying, and male ravens feed females and nestlings early in the nestling period (Ratcliffe 1997). Male ravens are larger than females, so what makes raptors different? Compared with similar-sized birds like ravens with straighter beaks and weaker talons, the hooked beaks and deeply curved talons of raptors offer several *advantages*, including the abilities to:

1 kill comparatively large prey;
2 dismantle large prey; and
3 carry relatively large prey balanced under the raptor instead of in the beak.

For example, male Southern Boobooks weighing ~270 g routinely kill and carry prey such as adult Crimson Rosellas weighing ~135 g. In contrast, slightly heavier (than a Boobook) Australian Magpies or similar-weight Pied Currawongs would have trouble killing, dismantling and carrying such prey to their nestlings. Vertebrate prey is rare in the breeding diets of magpies and currawongs compared with the breeding diets of similar-sized Boobooks, Australian Hobbies or Collared Sparrowhawks.

However, the hooked beaks and curved talons that characterise raptors have certain *disadvantages*. These include:

1 Large prey, such as birds or mammals, are often rarer in a bird's home range than are smaller taxa such as insects or spiders (Eltonian Pyramid; Krebs 2008), so raptors need to travel further to find vertebrate prey, and therefore use more energy to find vertebrate prey compared to finding invertebrate prey (Storer 1966; Reynolds 1972; Mendelsohn 1986).
2 Meat spoils quickly, so it is difficult to store for long periods in hot weather.
3 Because raptors travel greater distances to find larger prey, they leave nestlings and eggs unprotected from enemies and inclement weather for longer periods compared to birds that forage closer to the nest.
4 Having curved beaks and talons, raptors can only catch and carry one prey item at a time, so are energetically inefficient when capturing and carrying smaller invertebrate prey. If birds with straighter beaks, such as Eurasian Blackbirds, magpies or currawongs find an abundant source of crickets, spiders or worms, they can load several individuals into their beaks and carry these to the nest, as can Puffins *Fratercula arctica* (Figure 12.2). In contrast, raptors can only catch flying prey in their feet (not in their beaks) and carry one terrestrial prey such as a cricket, spider or worm to the nest. For example, breeding Little

Figure 12.2 Unlike raptors, birds without hooked beaks, such as the Puffin, routinely carry several small items in their beaks.

Owls in Eurasia take many worms (König and Weick 2008), but they only carry one worm at a time to their nest (Iñigo Zuberogoitia, pers. comm.). Herons (*Ardea* spp.) might be able to bring several fish at a time to nestlings, but Ospreys usually carry only one. The only published exception I could find was a report by Ellis and Nelson (2010) of Bald Eagles grabbing numbers of fish fingerlings north of the Queen Charlotte Islands in Canada, behaviour that the authors considered unique and uncharacteristic.

In Australia, all four *Ninox* owls, the Southern Boobook (Figure 12.3), Rufous Owl, Powerful Owl and Barking Owl, take flying or arboreal insects near their nests, and carry these invertebrates one at a time to nestlings; they travel further to capture and carry vertebrate prey (Hollands 2008; Stanton 2011; Olsen 2011a). This sets them apart from other owls weighing 300+ g around the world, none of which are known to regularly take aerial insects and deliver these to the nestlings (Jeffrey Marks and Iñigo Zuberogoitia, pers. comm.). It may be that curved beaks and talons incur *benefits* to insectivorous owls,

Figure 12.3 Fledgling Southern Boobook learning to hunt in the eucalypt canopy.

allowing them to easily dismantle hard-shelled invertebrates such as beetles, but incur *costs* when transporting such prey, so owls hunt them close to the nest and do not transport them long distances.

5 Curved beaks and talons allow raptors to capture more vertebrate prey but, unlike invertebrate prey, larger prey species such as birds or mammals tactically use cover to evade capture. This puts the pursuing raptor at increased risk of collision or injury. A raptor with even a slightly injured wing caused by hitting a branch or colliding with flocking birds will be unable to forage and will die. In contrast to the claim that female raptors need to avoid collision because of developing eggs (Walter 1979), *males* in particular need to reduce the risk of collision because they hunt more vertebrates than do females. Males need to reduce the number of dangerous hunting forays (Olsen 1989).

The differences in male–female roles in breeding raptors reflect at least three trade-offs related to costs and benefits of hooked beaks and curved talons:

1 **Energetics** – foraging for one large prey item at a time distant from the nest, *v.* catching smaller prey items close to the nest.
2 **Food storage** – buffer against irregular food (days with no food) and **protecting the nest contents** by remaining inactive near the nest *v.* searching for irregular food sources away from the nest.

3 **Risk to adults from collision, other injury, or predation** – pursuing larger, dangerous prey *v.* pursuing smaller, safer invertebrate prey.

These costs and benefits may drive the direction and degree of RSD in raptors.

Energetics (efficient foraging)

There is considerable evidence to support the prediction that smaller males in some species are more energy-efficient foragers than their larger female partners (Balgooyen 1976; Mendelsohn 1986; Storer 1966; Reynolds 1972). For example, Mosher and Matray (1974) examined the energy requirements of Broad-winged Hawks during the nesting season to test the hypothesis that a pair with RSD used less energy than a hypothetical monomorphic pair. They found no significant difference between male and female metabolic rates per gram for captive birds, that is, heavier birds – females at the same level of activity – used more energy in direct proportion to their weight. Because flight energy was estimated to be ~10 times more expensive than resting energy, the role division between male and female delivered ~20% savings on energy use during the breeding season when males were active for about four hours per day and females remained relatively inactive.

These energetic savings were part of the early models explaining RSD proposed by Storer (1966) and Reynolds (1972), but these authors stressed niche separation with male accipiters taking, on average, smaller, more abundant prey than females did (Eltonian Pyramid again). This seems to hold for some other raptors, for example Eurasian Sparrowhawks *Accipiter nisus* (Newton 1986), but not all. For example, Balgooyen (1976) found no difference in size of prey taken by male and female American Kestrels, and his re-analysis of Storer's data showed no difference in the size of prey taken by male and female accipiters, as Storer had claimed. Sunde *et al.* (2003) showed that breeding male and female Tawny Owls also took similar-sized prey, but males travelled further for food than females did, suggesting that selection was for males to be lightweight, energetically efficient foragers rather than specialists in smaller prey. There are many examples of this. Male and female Peregrines hunt the same small seabird on Langara Island, Canada, the crepuscular and nocturnal Ancient Murrelet (Nelson 1977). Furthermore, in certain raptor species including Southern Boobooks, Mississippi Kites and American Kestrels, males take *larger* vertebrate prey further from the nest and females hunt *smaller* prey close to the nest (Balgooyen 1989; Glinski and Ohmart 1983; Botehlo *et al.* 1993; Bader and Bednarz 2011; Liébana *et. al.* 2009; Olsen 2011a, 2013). Foraging patterns in these raptors represent a trade-off between the energetics argument and nest defence argument, linked to costs and benefits of curved beaks and talons, rather than the niche separation model (Figure 12.4).

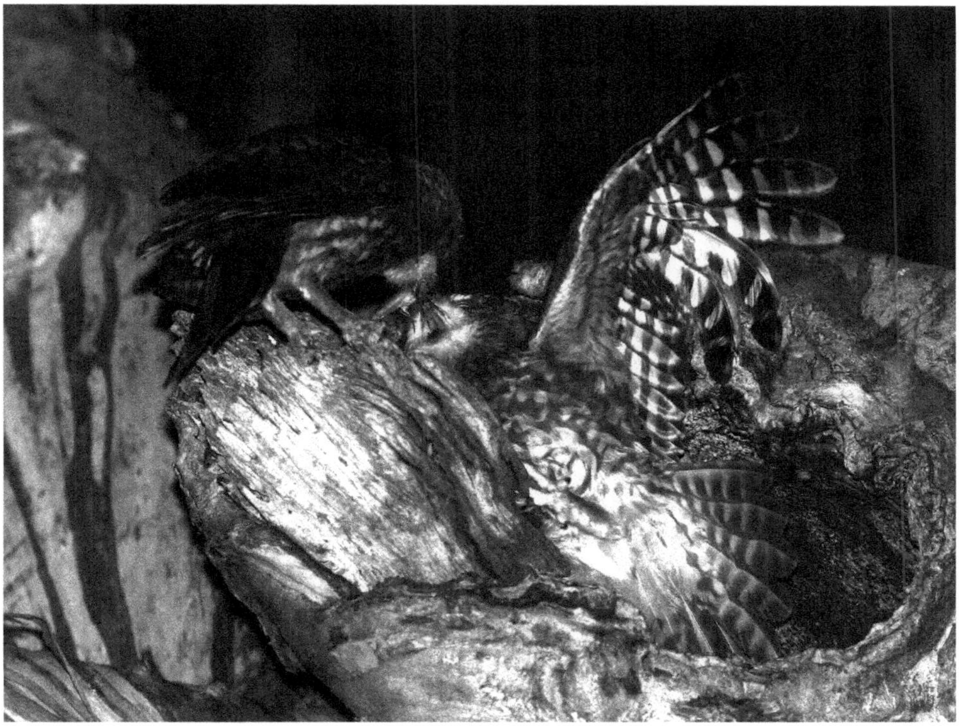

Figure 12.4 In Southern Boobooks, as with most raptors, the smaller male feeds his larger mate.

Buffer against irregular food and protecting the nest contents v. searching for irregular food sources away from the nest

A raptor's foraging patterns reflect other trade-offs, for example, females buffering against irregular food supply – days without food – by remaining inactive, protecting the nest, rather than searching for irregular food sources. Several studies support this proposition (Balgooyen 1976; Newton 1979; Mendelsohn 1986). If meat spoils quickly, and males cannot add weight to their bodies without increasing their risk of collision and decreasing foraging efficiency, it is better that females rather than males store that energy as fat. Females can also buffer against food shortage by hunting more. Researchers develop experiments to test such hypotheses around food shortage, by artificially increasing the number of nestlings (adding chicks from one nest to another), or decreasing brood size (removing chicks from a brood so adults do not have to hunt for them). Female flight time increased when broods were increased in Eurasian Kestrels, and female hunting ceased almost completely in reduced broods (Dijkstra *et al.* 1990). Female American Kestrels spent more time hunting when broods were enlarged in a poor vole year, but there was no clear pattern in a year of higher vole abundance (Gard and Bird 1990). In one of the most elegant studies, four nests of Great Horned Owls that originally held two young each were manipulated during high densities of cyclic Snowshoe Hare *Lepus americanus* populations in south-western Yukon, Canada (Rohner and Smith 1996). Broods of two nestlings older than 35 days were temporarily enlarged by one young, and then by two young. Radio-tagged parent females with enlarged broods did not respond to the addition of one young, but did respond to the addition of two young by moving further from their nests at night, apparently increasing their hunting effort. When extra food was supplied to the broods enlarged by two, the females' behaviour returned to normal, and they stayed closer to the nest and guarded the young. Massemin *et al.* (2000) found that smaller female Eurasian Kestrels in low vole years fledged more young than did larger females because smaller females *hunted* more.

There are also examples where females responded by *not hunting*. Female Ospreys in pairs with artificially increased broods did not hunt, but these sedentary females lost three times as much weight as their actively foraging male partners (Poole 1989). The females probably did not hunt because of the risk of nest predation by species such as Bald Eagles (Alan Poole, pers. comm.). Olsen and Tucker (2003) and Olsen (2011a) argued that field researchers may change female raptor behaviour by repeated visits to a nest, for example, to measure nestlings, so females become more reluctant to leave broods unprotected and refuse to hunt because of perceived risk of predation.

Risk to adults from collision, other injury, or predation

As discussed previously, hunting raptors are at risk of collision. Though owls such as Great Grey Owls *Strix nebulosa* and Short-eared Owls *Asio flammeus* take similar prey (mainly rodents), the forest-dwelling Great Gray Owl is far more dimorphic than the grassland-dwelling Short-eared Owl (König and Weick 2008). This may be both because the male Great Gray Owl risks collision while dodging between trees to search for and capture prey and due to the energetic costs of carrying prey around and between trees. Australia's two forest-dwelling *Tyto* owls, the Masked and Sooty Owls, are perhaps the most dimorphic owls in the world (Higgins 1999; Krüger 2005), but the grassland-dwelling Barn Owl and Grass Owl are among the least dimorphic. As expected, swift-flying falcons or woodland accipiters end up in rehabilitation centres out of proportion to their numbers in the wild (Olsen 1989).

Ninox owls with 'normal' sexual dimorphism

Unlike other owls, Rufous, Barking and Powerful Owls have 'Normal' Sexual Dimorphism (NSD). These owls take abundant prey in the forest canopy at two size extremes in relation to their bodyweight: 1) *heavy prey* – large and medium-sized, slow-moving, possums and gliders, and fruit bats, roosting birds; and 2) *light prey* – flying and arboreal invertebrates. Rufous, Barking and Powerful Owls seldom hunt the more intermediate-sized terrestrial mammals (Hollands 2008; Stanton 2011; Olsen 2011a,b). In contrast to Australian *Tyto* species, Australian *Ninox* do not have asymmetrical hearing so they are less effective at exploiting prey on the dark forest floor. *Ninox* do listen for prey, but they hunt mainly by sight. Females often hunt abundant insect prey near the nest; males often hunt arboreal mammals away from the nest.

In most other raptor species studied, males arriving with prey either give this prey to females, which feed or cache it, or males cache it themselves (Cade 1982; Olsen 2011a,b). In contrast, male Powerful Owls often roost on prey for the day (Figure 12.5) (McNabb 1996); females roost on prey after males relinquish it. This

Figure 12.5 *Ninox* species such as Powerful Owls have males larger than females and, unlike species with RSD, they roost on prey for the day.

could be read as an 'honest signal' (Simmons 1988), as well as being a way for larger males to cut their energetic costs and their risk of accidents by cutting the number of hunting forays (Olsen *et al.* 2013c). Males, then, are the main source of food storage in the Powerful, Rufous and Barking Owl pairs.

Relative abundances of arboreal marsupials and NSD in Powerful Owls

Small animals are usually more abundant than large animals (Newton 1998; Krebs 2008) and ecologists can predict abundance of an animal from its weight. For North American mammals, Krebs calculated a density of 60 individuals/km^2 for squirrels *Tamiasciurus* sp. (200 g) compared with only 0.5 individual/km^2 for Elk *Cervus elaphus* (250 kg) (Krebs 2008). However, it is not clear whether this relationship holds for arboreal marsupials, the main prey of Powerful Owls in eucalypt forests in south-eastern Australia. Could the relative abundance of their main prey explain NSD in some *Ninox* owls? Powerful Owls select large arboreal marsupials such as Common Ringtail Possum *Pseudocheirus peregrinus*, Common Brushtail Possum *Trichosurus vulpecula* (Figure 12.6) and Greater Glider *Petauroides volans* as their main prey (Fleay 1968) more often than they select smaller arboreal marsupials such as Squirrel Glider *Petaurus norfolcensis*, Yellow-bellied Glider *P. australis*, Sugar Glider *P. breviceps* and Feathertail Glider *Acrobates pygmaeus* as their principal food source (see e.g. Kavanagh 2002a). If these smaller arboreal marsupials are more abundant than larger ones, Powerful Owls are taking proportionally fewer of them compared to larger arboreal prey in the forest. However, if the opposite is true, i.e. larger arboreal marsupials are more abundant than smaller ones (*cf.* the normal Eltonian Pyramid), Powerful Owl prey selection is simply proportional to populations. This is suggested by several researchers, including Seebeck (1976), Cooke *et al.* (2006), Tilley (1982) and Lavazanian *et al.* (1994).

Common procedures for assessing the abundance of arboreal marsupials involve systematic spotlighting, transects, and/or listening for calls (Lindenmayer *et al.* 1999; Cooke *et al.* 2006; Wintle *et al.* 2005). Spotlight data of Cooke *et al.* (2006) and Tilley (1982) showed that larger arboreal marsupials were more abundant than smaller ones in the areas where they studied Powerful Owls. Table 12.1 contains published data from Kavanagh and Bamkin (1995), Bennett *et al.* (1991), Ward (2000), Tilley (1982), James (1980), Lindenmayer *et al.* (1999), and unpublished data from spotlighting counts conducted in Namadgi National Park and Tidbinbilla Nature Reserve by the Australian Capital Territory Parks and Wildlife Service from 1974 to 2009 (Murray Evans, pers. comm.) of counts of *smaller* arboreal marsupials (averaging 10 g to 600 g – Sugar Gliders, Feathertail Gliders, Yellow-bellied Gliders) versus *larger* arboreal marsupial mammals

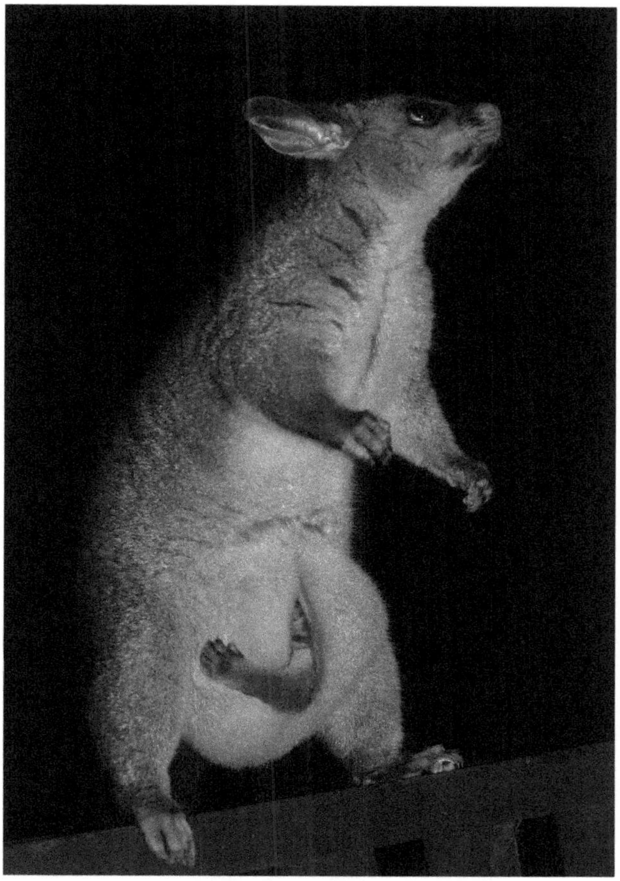

Figure 12.6 Common Brushtail Possums are at the heavy end of the spectrum of marsupials taken by Powerful Owls, but owls often take juveniles after they leave the pouch and ride on the mother's back.

(averaging 600 g to 4000 g – Common Ringtail Possums, Greater Gliders, Common Brushtail Possums, Mountain Brushtail Possums *Trichosurus caninus*) in south-eastern Australian forests.

Larger arboreal mammals were detected more often than smaller ones, with the difference between the frequency of small versus large highly significant ($\chi 2 = 1822.5$; $P < 0.0001$). Arboreal marsupials such as Common Ringtail Possums, Brushtail Possums and Greater Gliders were more common than smaller species such as Sugar Gliders (Figure 12.7) and Feathertail Gliders, and more common in the diets of Powerful Owls. This suggests that Powerful Owls are indeed taking larger arboreal marsupials in proportion to their abundance, though this needs further study.

Compared to most owls, Powerful Owls have proportionally small heads, as do forest-dwelling accipiters such as goshawks and sparrowhawks. This may relate to

Table 12.1 Published spotlight counts for eight species of arboreal marsupials in seven studies

Approximate mass of each species based on Van Dyck and Strahan (2008) and Menkhorst and Knight (2001).

Species and approximate mass	Kavanagh and Bamkin 1995	Bennett *et al.* 1991	Ward 2000	Tilley 1982	James 1980	ACT 1974–2009 unpubl.	Lindenmayer *et al.* 1999	Total
Mountain Brushtail 4000 g	**3.8%** *n* = 11	**6.2%** *n* = 87	**0**	**0**	**0**	**0**	**10%** *n* = 4	**3.3%** *n* = 102
Common Brushtail 3500 g	**4.2%** *n* = 12	**34%** *n* = 477	**4.8%** *n* = 3	**21.6%** *n* = 126	**11.3%** *n* = 6	**28%** *n* = 182	**25%** *n* = 10	**26.5%** *n* = 816
Greater Glider 1300 g	**24.5%** *n* = 70	**30.8%** *n* = 433	**59.7%** *n* = 37	**0%**	**22.6%** *n* = 12	**48.1%** *n* = 313	**38%** *n* = 15	**28.6%** *n* = 880
Common Ringtail 900 g	**4.5%** *n* = 13	**20.9%** *n* = 294	**30.6%** *n* = 19	**76%** *n* = 442	**30.2%** *n* = 16	**20.4%** *n* = 133	**27.5%** *n* = 11	**30.1%** *n* = 928
Yellow-bellied Glider 575 g	**23.4%** *n* = 67	**1.4%** *n* = 20	**0**	**0**	**0**	**0**	**0**	**2.8%** *n* = 87
Squirrel Glider 230 g	**0**	**1.5%** *n* = 21	**0**	**0**	**0**	**0**	**2.5%** *n* = 1	**0.7%** *n* = 22
Sugar Glider 125 g	**38.5%** *n* = 110	**4.2%** *n* = 59	**1.6%** *n* = 1	**2.2%** *n* = 13	**32.1%** *n* = 17	**2.9%** *n* = 19	**7.5%** *n* = 3	**7.2%** *n* = 222
Feathertail Glider 12 g	**0.3%** *n* = 1	**0.9%** *n* = 13	**1.6%** *n* = 1	**0.1%** *n* = 1	**3.8%** *n* = 2	**0.6%** *n* = 4	**7.5%** *n* = 3	**0.8%** *n* = 25
Total no.	**284**	**1404**	**61**	**582**	**53**	**651**	**47**	**3082**

the Powerful Owl's accipiter-like pursuit of quarry in the forest canopy, rather than ground prey. With small heads, Powerful Owls lack the spacing and asymmetrical ears found in Masked and Sooty Owls with broad heads that allow triangulation of sounds on terrestrial rodents such as rats, and marsupials such as bandicoots.

Though Common Ringtail Possums (Figure 12.8) are the most common prey reported in many dietary studies of Powerful Owls, one would assume that breeding pairs cannot survive solely on Ringtails. For example, if Common Ringtail Possums are an optimal prey size for both Powerful Owls and Sooty Owls, Powerful Owls would have to take prey the size of Ringtail Possums and *larger.* Sooty Owls, with RSD, could exploit prey the size of Ringtail Possums and *smaller* because they can take terrestrial mammals which probably fit an Eltonian Pyramid in abundance. It is interesting to speculate about why Australia has no quick-moving arboreal rodent species such as squirrels *Sciurus*. Perhaps eucalypt-feeders are comparatively slow moving because of problems with digestion, and slow-moving diurnals would be easy prey for diurnal raptors.

Figure 12.7 Sugar Glider.

Figure 12.8 Common Ringtail Possum (Julian Robinson).

13

Conserving high country raptors

In a biological sense, *conservation* means to preserve a wide variety of species, subspecies and habitats (retain *biodiversity*) and to save the greatest number of each species appropriate to the ecological conditions in any given area (retain *densities*). If the high country loses more forest, prey species, nest sites, open spaces or bushland, raptor species (biodiversity) may diminish. If houses, farms, mines, ski runs, wind farms or roads replace habitat that previously held food, shelter or nest sites, raptor densities will suffer.

Using the criteria devised by the IUCN (International Union for the Conservation of Nature) for their 'Red List', Garnett *et al.* (2011) listed all breeding raptors discussed in this book (high country ACT, NSW and Victoria) as nationally *Least Concern*. However, in the ACT and New South Wales the Little Eagle is listed as Vulnerable (i.e. facing a high risk of extinction in the wild in the medium-term future), with advice from experts to uplist this to *Endangered* (facing a high risk of extinction in the wild in the near future), and White-bellied Sea-Eagles are also listed as Vulnerable in Victoria.

Conservation techniques

Conservation techniques can be classed as **Preventive** or **Restorative**. Preventive techniques keep populations and habitats intact, before numbers are critically low. In this set of techniques, raptors are retained in habitats close to those in which they evolved. Preventive techniques stop problems before the species or subspecies becomes seriously threatened, or handle problems when taxa show early signs of threat but have not seriously declined. By contrast, restorative techniques handle problems after a taxon has significantly declined and is at risk of being lost.

Preventive techniques

Reserves – As a general rule, raptors require space to feel secure. They rarely cope well with close human interference, and they rarely nest close to houses. Although city parks and botanic gardens can offer abundant prey, vegetation and security from human disturbance for roosting and hunting, they seldom offer enough space and security for breeding. In contrast, national parks and reserves can offer abundant prey, suitable vegetation and security from human disturbance, and national parks and reserves are a particular characteristic of the high country. Kavanagh (2002b) noted that New South Wales and Victoria were the only states that had developed management plans on public forest lands (wood production forests and national parks) for large forest owls, and strong legislation in both states protect these owls. Kavanagh (2002b) outlines conservation protocols used in NSW wood production forests, for example, exclusion buffers of 10 to 40 m around rainforest, wetlands, heath, rocky outcrops, etc., and the retention of hollow-bearing trees and dead trees.

Three concepts need to be mentioned that can help a developer more than raptors. A **buffer** set around an active eagle nest is often of little use because the buffers are too small, and they don't consider the size of the whole home range and how eagles move nests sites as far as 5 km within a breeding home range. Buffers may stop some disturbance for a short time, but free developers to destroy foraging habitat and areas for alternative nests, so the eagles eventually abandon the area. **Corridors** set in the ACT, said to link areas of habitat, are usually too narrow to sustain pairs of eagles, so eagles will eventually be lost in these areas, unless adjoining farmland outside of the corridor is kept free from development. **Trade-offs**, where bodies claim to retain, for example, an area of box woodland in one place if they can destroy it in another place (Figure 13.1), are often a form of sleight-of-hand because both areas of woodland should be retained, not one. These trade-offs or offsets rarely trade like for like, and can't truly be offsets unless new habitat is created, which takes decades.

Most of the Australian Alps are now national park, but land-use decisions vary from park to park and region to region. The core of Namadgi National Park is the Cotter Catchment, primary source of domestic water for Canberra and Queanbeyan. The area has been protected since the establishment of Canberra in 1908; in fact, the shape of the ACT was chosen as the borders of this Cotter Catchment. This benefits raptors in the southern ACT, but species such as Little Eagles that breed predominantly in the north escape protection. One dictionary definition of conserve is to 'keep from harm, decay or loss, especially with view to

Figure 13.1 Prime raptor breeding habitat being destroyed in Canberra for the Gungahlin Drive Extension.

later use' (Turner 1987). All past uses of the high country have changed it. Indigenous peoples changed the high country by using fire to favour some plant species and multiply food sources, including animals, at the expense of others. Graziers introduced stock, and many would like to continue stock grazing in national parks. Land use is a compromise, an uneasy tension between vested interests: skiers, graziers, walkers, photographers and bikers all disturb animals and change the landscape. High country raptors need better study before a species or subspecies declines, an assessment of resources that these taxa need to retain biodiversity and densities. In general two types of research underpin sound, preventative techniques:

1 Research to uncover the causes of a species' or subspecies' vulnerability.
2 Research measuring which techniques most effectively conserve a species if it declines or becomes threatened.

Most raptor conservation problems are linked to increasing human populations and the high country is no exception. The relationship between changes in habitat and human population growth is crucial to determine which species and subspecies are harmed by increased human population, and which are benefited, and why.

Restorative techniques

Restorative techniques attempt to return raptor populations that have decreased to critical levels. These can involve captive breeding, transporting species to healthier habitats or rebuilding habitats, for example, with nest boxes (Figures 13.2, 13.3), to increase the numbers of breeding pairs. Normally this means that raptors are retained in dramatically altered environments from those the species evolved in, so it should never take precedent over saving good habitat, though sometimes it does.

Figure 13.2 Nankeen Kestrel returning to a nest box.

Research

Fundamental research on threatened and vulnerable species and subspecies underpins techniques. The 'tools' include habitat maintenance, nests and shelter, food, captive breeding and rehabilitation (Appendix 2), and fostering (Figure 11.13).

Habitat maintenance

The most critical measure for conserving raptors is maintaining their residual habitat. Two measures are necessary: 1) find the remaining areas of good habitat and preserve them; and 2) increase the carrying capacity of these areas so they support more raptors. These remnant areas should provide secure nest sites, shelter, food, space and freedom from disturbance.

a) **Nests and shelter** – In many places there is ample food for raptors but too few suitable nest sites and little shelter. Owls and falcons will often accept a man-made structure that resembles their natural nest. Nest boxes and artificial stick nests can play an important role in species and population conservation as long as the chosen sites are protected from inclement weather, direct sunlight, predators and human interference. However, nest boxes and artificial stick nests are no substitute for maintaining natural nest sites in natural habitat; this usually means retaining natural tree hollows and woodland.

b) **Food** – Availability of food and nest sites sets the carrying capacity of land containing raptors (Newton 1979). Usually it is harder to raise the carrying capacity of a piece of land by increasing food supply than by increasing the number of available nest sites. Increasing food supply usually entails changing land use or management to promote an increase in prey numbers. Again, this is best achieved by preserving good habitat in the first place, on farms, in cities, reserves and in national parks.

Biologists can assess habitat quality for prey, including introduced species such as European Rabbits, Common Starlings, Common Mynas, Black Rats, House Mice and House Sparrows that can benefit some Australian raptors. Fortunately, introduced prey species in the high country were less contaminated by organochlorine pesticides than in other parts of the world, although rodenticides may be a new problem.

Habitat restoration and corridors

Debus (2009) noted that restoration of some indigenous tree cover on formerly wooded rural land is a necessary conservation technique because the large owls may be reluctant to traverse extensive open country. Corridors planned by some city governments are too small to accommodate large breeding raptors, especially Wedge-tailed Eagles.

Habitat loss – Habitat is degraded in three ways: 1) the raptor can be left with tiny fragments of habitat, for example, small patches of rainforest; 2) the raptor can be left with a general alteration of its former habitat, for example, logging or the farming of Australian grasslands. Both these forms of degradation can affect biodiversity, but where fragments are left, raptors can be found within these fragments at no less a density than before. However, where general habitat degradation occurs, the prey used by raptors may be reduced. The raptor species or subspecies still occurs over its original range but at greatly reduced densities; and 3) habitat can be totally destroyed and fragments are so small that raptors can no longer live there.

Timber harvesting

All breeding high country owl species are woodland or forest species, as are the three accipiters, so it is important to single out timber harvesting as a special threat. Kavanagh (2002b) noted that large forest owls occur at similar frequencies in unlogged and selectively logged areas, but they occur at lower frequencies in heavily logged areas. Where unlogged reserves were retained along streams and smaller drainage lines where Common Ringtail Possums occurred, the impact of logging on breeding owls was greatly reduced. Clearly there is more we need to understand about the effects of logging and owl populations, and no such studies have been done with Australian high country accipiters. Some owls and accipiters are successfully moving into major population centres, so the prognosis looks promising in part.

Wind farms

Wind farms are an increasing threat to high country raptors, especially eagles near Lake George. Wind farms are also planned for the ACT. Studies are needed, both for the impact of wind farm mortalities on raptors, and techniques for mitigating these impacts.

Using raptors to conserve non-raptor species

Raptors are commonly used as 'surrogate species' to help conserve plants and animals at lower trophic levels. The argument goes that saving habitat needed by eagles, hawks, falcons or owls in order to save less 'glamorous' species of bird, tree, grass, reptile and insect benefits conservation. Top predators need space and, through predation, they affect other animals and plants at lower trophic levels, so conservationists can save many plant and animal species at lower trophic levels by saving raptor species. Furthermore, some raptors look 'glamorous' compared with other animals, so they can be used as conservation tools to draw attention to vulnerable species and areas, and focus government or public attention on areas or taxa that would otherwise be difficult to conserve. Hence, the raptor is a 'surrogate', a 'substitute' for large, costly studies of biodiversity. Saving top predators saves species richness.

Raptors can be used as 'surrogates' in at least five ways, as 1) Biodiversity indicators; 2) Keystone species; 3) Umbrella species; 4) Sentinel species; or 5) Flagship species (Caro and O'Doherty 1999):

1 **Biodiversity indicators** – Areas with breeding raptors are said to contain more biodiversity (species richness) than same-sized areas without breeding raptors. Conserving areas with raptors will conserve several other species.
2 **Keystone species** – The raptor has a disproportionate impact on the environment relative to its abundance because of its predatory habits. For example, Eagle Owls in the European Alps might depress the number of Tawny Owls in an area, and high numbers of Tawny Owls depress the number of Scops Owls, so keeping Eagle Owl numbers high keeps Tawny Owl numbers low, and this keeps Scops Owl numbers high. At the moment, there is little evidence in the literature to support this notion. It applies to certain circumstances but not others, and it can be difficult to predict accurately how a predator will affect species at lower trophic levels in its home range. Ecosystems are complex.
3 **Umbrella species** – Raptors need large home ranges to breed and hunt in, so saving raptor home ranges also saves animals and plants inside these large home ranges, including the smaller home ranges used by smaller species. In contrast to using raptors as Biodiversity Indicators and Keystone Species (above) where researchers select a certain area as a good place for a reserve because it contains breeding raptors, and raptors affect biodiversity through predation, this measure is about *size*. Umbrella Species provide large areas with more biodiversity (species richness) not so much because they have more species than other areas of the same size, but simply because they are large. There has been little quantitative investigation of the effectiveness of top predators as Umbrella Species, but the umbrella role may be useful to help us 'think large', thus promoting larger reserve size (Ray *et al.* 2005).
4 **Sentinel species** – Raptors are sensitive to human-induced changes in the environment, so studies of breeding raptors can detect conditions in the environment before these changes affect other species. Raptors are said to be reliable indicators of the condition of the environment because of their position at the top of the food web, and because they live at low densities, have low fecundity, have long periods of juvenile dependence, and are the first to disappear when habitat is degraded. This makes them more vulnerable to human-induced changes in ecosystems than are some other animals. The best-known example is the research on toxic pollutants like DDT and PCBs that

triggered extensive investigations in the 1950s and 1960s because of their effects on bird- and fish-eating raptors. Derek Ratcliffe (Chapter 10 above) found eggshell thinning in Peregrine Falcons, and his research was used to stop pesticide use in North America, Europe and Australia. More recently, owls and perhaps eagles have signalled problems with rodenticides in the environment (see Chapter 9).

5 **Flagship species** – Raptors, especially Wedge-tailed Eagles, Peregrines and Powerful Owls, can be charismatic poster species. Conservation organisations use owls as logos (for example, the Australasian Raptor Association) on magazine covers, and in advertisements to attract funding and focus public attention on larger causes. In the 1960s the Peregrine Falcon was not only a Sentinel Species (see above) but also an icon of conservation campaigns to ban pesticides.

Besides economic gains, Flagship Species are ideal tools for public education. Charismatic species capture public attention more effectively than do abstract concepts such as biodiversity or ecosystem integrity.

Raptors work as Flagship Species. However, some might argue that conservation already concentrates too much time and money on charismatic species, and conservation needs to be driven, instead, by ecological need. Some animals attract millions of dollars of public and government funding simply because they are charismatic, not because they are rare. The argument is that conservation should be directed by research findings, not by public fashion. In addition, Australia lost two Boobook subspecies in Australia at the same time that considerable money and effort was being spent researching Wedge-tailed Eagles, Powerful Owls and Peregrine Falcons, charismatic species that were not threatened. Hopefully we will not make this mistake a second time. Wedge-tailed Eagles, Powerful Owls and Peregrine Falcons are not endangered in Australia, but we have almost no information on the status of island raptors species and subspecies in Australia and the region.

As surrogate species, raptors can lever public support, raise funds, preserve biodiversity, protect and prioritise reserve sites, and help in the planning, size and configuration of protected areas (Gittleman *et al.* 2001; Ray 2005). But we need further research to validate the value of raptors as Umbrella Species, Keystone Species and Biodiversity Indicators, and the potentially adverse affects of concentrating on secure continental species to the detriment of island forms.

Canberra raptor communities

Frith identified Canberra as the major metropolitan area in the high country, so it is important to look at conservation issues there. Canberra has breeding eagle species inside its city limits as well as kites, owls, hawks and falcons. The guild of raptors in Canberra is unique, and indicates a healthy number and variety of prey animals at lower trophic levels. It is crucial to retain as many nests as possible, and retain all species currently breeding in the Canberra guild, in line with the policies laid out by the ACT Land and Planning Authority (2004, p. 15) to 'better protect the biodiversity within the ACT and regional ecological communities'.

Raptors that breed successfully in urban environments

Two raptor species breed successfully in the central city. Collared Sparrowhawks and Australian Hobbies breed and hunt near some human activity – walkers, automobile traffic, even sporting events, close to or even under the nest. They can successfully breed in city parks, even parking lots, that is, in city landscapes outside of Canberra's Nature Parks and Reserves. No other ACT raptor species does so, though Peregrines and Kestrels could be induced to breed on city buildings as they do in Melbourne, Sydney, Brisbane and many world cities.

Raptors less likely to breed successfully in urban environments

Black-shouldered Kite – Pairs in the ACT seem to require several hectares of tall grassland per pair to survive and breed. If houses and suburbs expand to cover large grassland areas, roads or playing fields, the species is likely to disappear from those areas.

Whistling Kite – In the ACT these kites require large areas of aquatic habitat for foraging. Only one pair was found breeding in 2009 and none in 2011, and the species may become extinct in the ACT because of urban development around Kingston, and possibly because of poisoning of pests such as rabbits.

Swamp Harrier – In the ACT these harriers require large areas of aquatic habitat and grassland for foraging. Only one pair was found breeding in 2009, in Namadgi National Park, and the species could become extinct in the ACT. This species is the most sensitive ACT raptor and will abandon its nest and young if disturbed by humans.

White-bellied Sea-Eagle – This species is listed as Rare in the ACT. The species has not been confirmed breeding in the ACT but non-breeding 'floaters' occur regularly along the Molonglo River and these birds are likely to leave the Molonglo Valley if the area is urbanised.

Wedge-tailed Eagle – We have good data on food habits, but little data on foraging habits and habitat requirements of this large eagle. In the Canberra region, each pair of eagles was said to require 28 to 32 km^2 to breed, or at least this was the density recorded (Leopold and Wolfe 1970), but this probably is an underestimate. The species is highly susceptible to disturbance, and where they nest close to suburbs in the ACT, they usually use tracts of land seldom used by walkers. If people use the current nesting areas for recreation, and the macropods that make up the bulk of their diet falls in number, some pairs are likely to disappear. A buffer zone of at least 500 m radius around the nest site with no walking tracks or other human activity, especially from April to December of each year, will help, but a problem with this species is that alternative nests can be as much as 5 km apart. A buffer around last year's nest will not help if the pair has moved.

Wind turbines built near Collector are likely to chop up Wedge-tailed Eagles and probably Little Eagles and other species.

Little Eagle – Though several pairs nested in the ACT in the 1980s and 1990s (Mallinson *et al.* 1990; Olsen 1992), many of these disappeared because of urban development, and maybe because of secondary poisoning from Pindone and other chemicals used to control rabbits. Other reasons for the decline, including crowding, disease and displacement by Wedge-tailed Eagles should be investigated. Though listed as Vulnerable, Little Eagles should be re-listed as Endangered in the ACT before they disappear as a breeding species.

Brown Goshawk – This species prefers stands of eucalypts, especially Red Stringybark and Red Box woodland for nesting, and if these disappear, the goshawks are likely to disappear. Some pairs are disturbed by walkers and leave their eggs and young exposed to die.

Peregrine – ACT pairs have nested on the same cliffs since at least 1976, and because they require cliffs to breed, these areas are critical. Peregrines forage for birds, especially parrots, over grassland and river habitat, though it is not clear how far they go in search of prey. There are no telemetry studies for Australian Peregrines. We believe that, for successful nesting, each cliff needs a buffer zone of at least 200 m radius around the nest site with no walking tracks or other human activity, especially from August to December of each year. At present, the ACT government plans to fill in and destroy the quarry in which Peregrines nest on Mt Mugga (Figure 2.5). Perhaps the falcons can be induced to nest on nearby buildings in Woden (Figure 13.3).

Brown Falcon – This species uses abandoned nests, often raven nests, in mature eucalypts. They like the shores of the Molonglo and Murrumbidgee River Corridor less than some other species, and tend to nest higher up on hillsides in farmland away from the river, but they do nest along the rivers in several places. They eat a wide variety of prey, and there is considerable variation between the diet of even neighbouring pairs. They require large areas for foraging, and suburban development has already caused the abandonment of some territories. We need telemetry studies to better understand their foraging behaviour and territory size. Generally, they do not nest close to houses in the ACT, and are likely to abandon nests if there is too much human activity near their nest.

Nankeen Kestrel – This species lives in cities, roosts on buildings, and could possibly use nest boxes. However, most require short grasslands for foraging on small dragon lizards, birds and insects and it is not clear whether Canberra holds

Figure 13.3 Peregrine nest box attached to a wall in the city (Paul Willcock).

enough prey for breeding. Outside the city they require mature hollow trees for nests. If the short grasslands used for foraging are developed for housing and roads, breeding pairs will disappear.

Southern Boobook – Nests are difficult to locate. They roost and nest in mature hollow trees, and where we studied radio-tagged birds on Black Mountain, Aranda Bushland and Bruce Ridge, the pairs nest in bushland, then hunt after dark in the same woodlands or cross into suburbs. Usually, they do not nest in suburbs, but occasionally they will use parks or bike paths if trees are big enough to provide suitable hollows, and elevate the nest cavity to a safe distance above the ground. Common Brushtail Possums are a danger as they take eggs and young,

Figure 13.4 Nestling Southern Boobook being ejected from its nest by a Common Brushtail Possum.

and usurp nest hollows. Culling dead trees removes nest sites and is especially destructive to populations of Boobooks and kestrels. If large tracts of woodland are left in the ACT with mature eucalypts with nest and roost hollows, and plenty of bushy trees such as *Exocarpos* for day-roosting, these owls should continue to breed in good numbers. Nest boxes protected from possums (Figure 13.4) and other predators could work in certain areas. Anticoagulant rodenticides may cause owl deaths in Australia.

Powerful Owl – Nests are difficult to locate. They roost and nest in mature hollow trees; nests found in Namadgi and Tidbinbilla were away from habitation. Before the 2003 fire we estimated eight pairs in the ACT, after the fire only one pair. They usually do not nest in suburbs. If large tracts of forest are left in the southern ACT with mature eucalypts with nest and roost hollows, and plenty of bushy trees, like *Exocarpos*, and arboreal marsupial prey, these owls could continue to breed in good numbers.

14

Giving care

An eagle was in trouble at the edge of the city. It stood on a busy road where rangers were anxious that it might be struck by a car, and later it sat in a low tree between the road and an active quarry full of dangerous machinery. The rangers thought it was injured, but they couldn't catch it, so we drove across town and watched it through binoculars. It looked like a young female, maybe injured, so Felicity Hatton and I approached her with a noose pole trap but the eagle flew and coursed low over a hillside towards the quarry. We rushed after, threading our way through eucalypts and long-stemmed yellow flowers, trying to keep the eagle away from the quarry by blocking her downhill flight-path, always forcing her to fly uphill. She landed in grass, spread her wings to catch the wind and was about to take off when I grabbed her.

Ten years earlier, on the isolated Indonesian island of Roti, Sue Trost and I wandered into a village one night and were told of a woman with a caged owl in one of the huts. The island was full of kind but very poor people, and a source of teenaged crew for people-smuggling boats. A teenaged boy took us to the woman. It turned out to be a nestling Brown Goshawk maybe two-thirds grown (Figure 14.1), in a wicker cage with two shrivelled fingerling fish; the hawk looked unhappy and hungry. Two boys told us they had shot down the nest with slingshots and captured the nestling as a gift for their mother. Throughout Indonesia people commonly eat nestling goshawks, kestrels, owls and other raptors, but this one was destined to be a pet that would surely die slowly.

The following morning we examined the nestling in better light, and saw no visible injuries or sickness. Villagers watched; a girl of eight or nine rested a naked

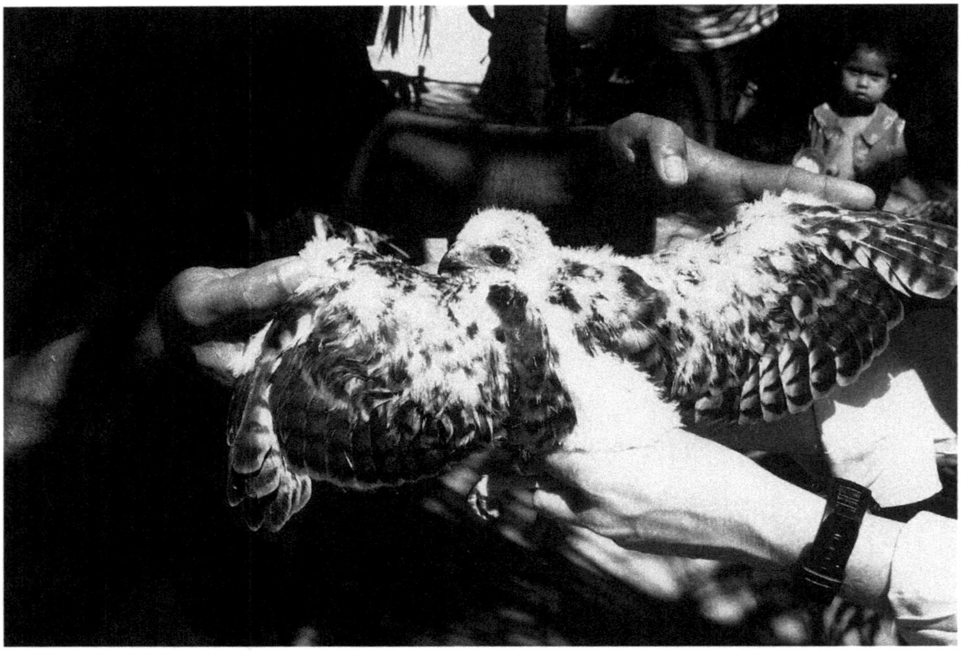

Figure 14.1 Nestling Brown Goshawk on Roti.

baby on her hip. We convinced the boys to show us the location of the nest tree, and then walked half a kilometre through forest in morning heat. The tree was along a stone fence surrounding a cornfield, and we tried to convince the boys to return the nestling. They said this was impossible because the nest now lay on the ground, which it did, and the nestling had been in the village for the better part of a week and, most important, the mother goshawk was gone. We convinced them to try anyway, and an older man from the village climbed the bare trunk with a basket full of sticks and leaves, sticking his big toe into axe cuts that he hacked into the trunk as he climbed. He tied the basket to a fork high on the tree, left the nestling in the basket with a dead day-old chick, and started down. At the bottom he hit the ground, smiled and raised his arms in a strongman pose. All the watching villagers laughed.

The nestling immediately began food-begging, a plaintive call to its parents, and the female goshawk came in with a dead passerine. She tore it up and fed it to the nestling. She was of the race *Accipiter fasciatus hellmayri,* white-headed and smaller than the Brown Goshawks in Canberra, but their vocalisations were identical. That week we left money for the villagers as payment for monitoring the goshawk's progress, and nominated a 'case manager' who reported back to us in Australia. The nestling, he reported, had fledged, hung around for 3 weeks being fed by both parents, then left.

We had done this many times in Australia. Most diurnal raptors don't recognise their own young so they feed strange young, or their own long-removed young. For example, we'd placed Wedge-tailed Eagles bred at Taronga Park Zoo into the nests of wild eagles in the high country, and parents had raised them as normal. If the downed eagle at the edge of the city was a juvenile, we knew what to do. She seemed big in our hands so we started calling her a female. We drove home and examined her from head to toes under electric light in the bathroom. There was no injury to her head or swelling, and her eyes dilated as normal in torchlight. There was no injury to her cere or nostrils nor any discharge, no fractures or splits to her beak, and her breathing was quiet with no gurgling (fluid) or panting. Her neck was uninjured, no food in her crop, and her keel (*sternum*) was a little sharp, so she was slightly low in condition, but there was no blood around the vent indicating internal injury, or matting (excreta) indicating worms, poor health, disease or, again, internal injury. There were no breaks or injuries to bones in her wings or legs. All primary and tail feathers were intact, and we rocked her gently from side to side to check the symmetry of her wings and legs. There were no bumblefoot lesions on the pads of her feet that would indicate she'd been in captivity and escaped. But the bases of her primary and tail feathers were blood-filled quills, telling us she'd recently fledged and drifted into danger.

Colin MacLaren had fashioned a hood that fitted her well and calmed her (Figure 14.2). She sat on a wooden box in the bathtub. There was a purple bloom

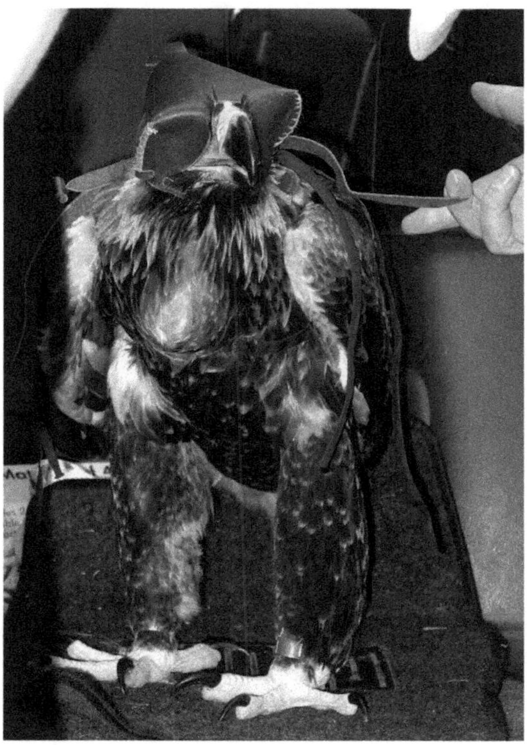

Figure 14.2 Hooded Wedge-tailed Eagle.

Figure 14.3 Radio-tagged Wedge-tailed Eagle (Mark Sweany).

off the new feathers on her back. To get her strength up, and give her a full crop when released, I fed her every 6 hours through the hood, as falconers did hundreds of years ago with newly caught eagles in Asia. This would help us adjust the radio-harness (Figure 14.3) below her full crop instead of across an empty crop. Occasionally in the night she ruffled her feathers, and shook her head with the buttons on the end of the hood laces tapping against leather. Her talons clicked on the wooden box as she stepped around in the dark.

As the villager did on Roti, we planned to climb a tree near her nest tree, give her a second chance, even though we knew first-year mortality in raptors was high. Why do this? Animal ecologists generally hold animal rehabilitation in disdain because it does little or nothing for animal populations. Saving the goshawk on Roti, or this just-fledged eagle, seems pointless. But most of us who study raptors genuinely love certain species, and feel a certain affection towards individuals we study, a drive to care for them. We want to conserve habitats and populations, and better the lives of individuals. So, many of us spend time helping individual raptors in trouble even if there is little ecological or publication value in it. At its heart is the basic question of all questions – how do we spend our limited time? There would be some ecological and publication value in attaching a radio to this eagle, but she would have been treated the same anyway.

We placed her up in an Apple Box close to her nest, hooked her feet onto a fat limb, removed the hood, and I slid down the tree before she flushed again. We turned and walked downhill and waited for her parents to come.

Appendix 1 Diurnals

Note: species accounts for owls can be found in the book *Australian High Country Owls* (Olsen 2011a).

(Julian Robinson)

Black-shouldered Kite

Elanus axillaris

Elanus is said to be a Greek word for 'kite'; *axillaris* refers to the striking black patches on the wings and is from Latham's *Axillaribus nigris.*

Field recognition

Signature characters: A little larger than a magpie or Nankeen Kestrel. **Adult** light grey back, white front, black shoulder patches and pointed black wingtips; appears overall white from a distance, pumps tail up and down when perched, hovers; juveniles browner. Black patch in front of bright red eyes and extending over them as an eyebrow. Cere horn to yellow, feet bright yellow. **Juvenile** has diffuse rusty-brown streaking and wash on the head, back and crop area. Scapulars and feathers on upperparts, including black wing patch, are edged in white; cere is horn coloured, eyes are brown and feet are pale yellow.

Perched, it looks legless, and wingtips extend beyond the tip of its relatively long tail, even though its tail is shorter than a kestrel's. Its flight is rapid and falcon-like, with winnowing wings. It glides on wings held in a dihedral, hunting typically in arcs and circles over grasslands, frequently hovers with legs lowered and tail depressed; descends leisurely at first, then head-first, then feet-first with wings raised above its back in a sharp V.

It is 33 to 37 cm long with a wingspan of 82 to 94 cm and slightly dimorphic but not evident at a distance; males average ~249 g, females ~293 g.

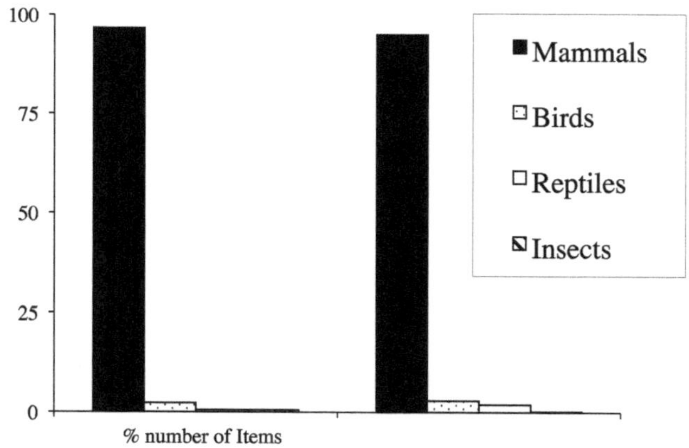

Figure A1.1 Diet of the Black-shouldered Kite around Canberra (Olsen, Fuentes and Rose unpublished).

Habitat: Diurnal and crepuscular in open areas with long grass dotted with mature trees, rare or absent from forested areas. In the highlands it is generally found in pairs or singly or in family groups and seen flying or perched on dead treetops. In the ACT they are most often seen on grasslands or disturbed habitats outside of the city, including post-fire pine plantations.

Habits: They hunt from a high perch or by quartering and hovering, which provides a stationary platform to scan the ground and drop vertically, parachute-like, into grass onto their prey. Very aggressive to other raptors and birds at which they dive and strike, especially while breeding. Minimum home-range size when breeding has been estimated as ~200 ha in the ACT (Barnes 2005) and northern inland NSW (Debus *et al.* 2006).

Food: Occasionally take small birds, lizards and insects, but in the high country they specialise almost solely on introduced rodents, ~96% House Mouse (Figure A1.1).

Breeding: May breed when one year old. The only high country raptor that breeds throughout the year (Figure A1.2), often in spring and autumn, so some pairs have two broods in a year. Continuous breeding has been reported in other parts of Australia when food is plentiful. The nest is small, 27 to 45 cm across and 13 to 15 cm deep, similar in size to a magpie or currawong nest, placed high in a live or dead eucalypt, acacia or pine, usually in an open paddock or at the edge of woodland.

Eggs: Two to five, rounded, creamy-coloured splashed brownish and reddish. Incubation said to be ~31 days.

Young: Nestlings are sandy or pale brownish in their first down, pale to light grey in the second down, slightly browner on the crown. Young fledge at ~33 to 38 days after hatching and may stay with the adults for 3 or 4 weeks. They can disperse long distances, up to 1000 km.

Mar April May June July Aug Sept Oct Nov Dec Jan Feb

(Spring breeding)

I—I
Pre-breeding
I—I
Eggs
I—I
Hatching
I————I
Nestlings
I—I
Fledglings

(Autumn breeding)

I—I
Pre-breeding
I—I
Eggs
I—I
Hatching
I—I
Nestlings
I—I
Fledglings

Figure A1.2 Approximate timing of breeding for Black-shouldered Kites in the high country.

Voice and calls: Wheezing whistle, harsh scraping *skair-skair.* Also, a soft, descending whistle, a contact call, and a short *chirp-chirp-chirp* or *kik-kik-kik* when defending against intruders or to mates.

Status: Not nationally or internationally threatened. It is widespread and common through much of its range, depending on prey availability. Burning, other land clearing and the introduction of the House Mouse may have benefited them.

Myths and questions: Although common, it is one of the least-known Australian raptors. Around Canberra, the species sometimes does not breed in certain years, perhaps because of food shortage, or because pairs have moved to food sources away from Canberra.

Sources: Olsen unpublished data; Marchant and Higgins (1993); Debus (2012); Hollands (1984); Olsen (1992); Olsen and Rehwinkel (1995); Barnes (2005); Read (2005); Debus *et al.* (2006).

(Julian Robinson)

Whistling Kite

Haliastur sphenurus

Haliastur is for 'sea-hawk', Greek *Hal* for 'salt, sea' and Latin *astur* for goshawk; *sphenurus* combines two Greek words for 'wedge' and 'tail'. 'Sea goshawk' denotes the often coastal nature of the genus, but 'wedge tail' is misleading as tail-tip is rounded. Originally it was called *Milvus sphenurus*, but recent DNA evidence has confirmed the distinction.

Field recognition

Signature characters: Loud whistling, down-sliding call of six or seven notes; **Adult** is sandy brown, larger than a raven, perches upright, bare tarsi, rounded tail-tip, fingered primaries in flight, deep jerky motion, slow, circling. Beak and cere are horn, iris dark brown, feet whitish-cream. **Juvenile** similar to the adult but darker and rustier with underparts more streaked, heavily spotted buff or white on back and underwings.

Medium-sized, scruffy, sandy-brown kite, slow flying, often circling on slightly bowed wings and long spread primaries. Large pale area on upper wings, a long tail with rounded tip, and dark outer flight feathers. It flies with a rather jerky rowing action accompanied by some body movement and glides on bowed wings, bare tarsi visible. At rest feathers can mostly cover its short legs leaving only toes showing; the long folded wings do not reach to the tail-tip.

The Whistling Kite is 51–59 cm long (tail about half its length) with a wingspan of 123–146 cm; Males weigh on average 632 g, female 907 g. Similar in size and appearance to the Little Eagle, so the two can be confused. Some Whistling Kites have a characteristic 'window' pattern in the wings, described in most field guides, though this characteristic is not as discriminating as some suggest. Little Eagles have a shorter, squarer tail. The Whistling Kite is less stocky, has a smaller head, no feathered legs, and glides on deeply bowed, not flat wings. The closely related Black Kite occasionally occurs in the high country, darker than Whistling Kites with a narrower, pale band on the upper wing, narrower, more pointed wings and a forked tail.

Habitat: In the high country Whistling Kites are seen mostly alone or in pairs on grasslands, marshes and open woodland close to the upper Molonglo River and Lake Burley Griffin north-west of Canberra, as well as the Murrumbidgee and other rivers. They do not occur in dense forest, but groups have been seen at Tidbinbilla.

Behaviour: Characteristic behaviour is lazy soaring, with fingers of the primaries extended; sometimes they give loud whistling calls as they fly or from prominent perches or the nest, standing in an upright position, tossing the head back, tilting the bill skywards. Whistling Kites perform aerial displays, mutual soaring, acrobatics and mock attacks. They allopreen and males present food to females at the nest.

Territoriality: Adults at the nest attack other raptors including other Whistling Kites. Behaviours include a loud whistle, prominent perching and soaring.

Food and hunting: Carrion but also live prey including rabbits and rats, reptiles including lizards and small snakes, birds, fish, crustaceans and insects as

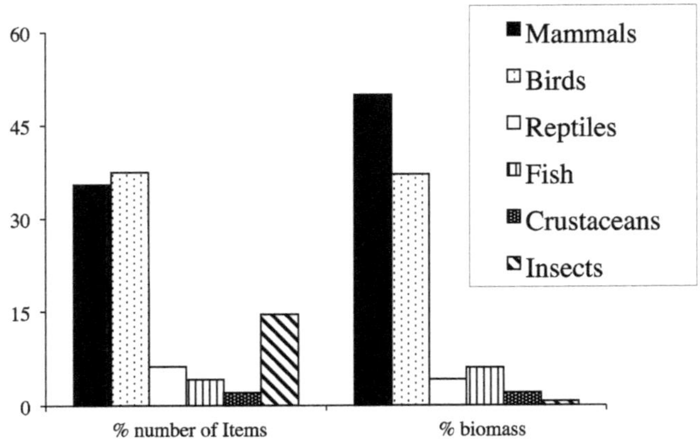

Figure A1.3 Diet of the Whistling Kite in the ACT (after Fuentes *et al.* 2005; and Olsen *et al.* 2013d).

well as road-killed vertebrates, large carcases and offal. In the ACT rabbits and birds of different sizes dominate the diet (Figure A1.3). They forage by high quartering, soaring and by still-hunting from trees. They land next to carrion and consume it in groups, kill prey on the ground, hawk flying insects, and snatch prey from the water. I have seen them make repeated stoops into flocks of Common Starlings. They forage along roadsides and fire fronts and steal from other raptors.

Breeding: No detailed studies. In northern Australia they can breed in all months; the laying season is long and variable, but on the Southern Tablelands they start breeding in winter and early spring (mainly March to October) (Figure A1.4). They may nest further from water than is usually reported, with nests located on open farms and on hillsides as far as 1.5 km from rivers. They nest alone, or in loose groups with pairs 100 m apart. The nest is a platform of sticks often larger than a raven's nest, 60–150 cm diameter, 30–100 cm deep, with the inner cup lined with green leaves, and placed 3–62 m above the ground in the fork of a live or dead tree. They breed around the east end of Lake Burley Griffin and west of the ACT on the Murrumbidgee River and Burrinjuck Dam, but not on the lower Molonglo River in recent years.

Mar April May June July Aug Sept Oct Nov Dec Jan Feb

I——I
Pre-breeding
I——I
Eggs
I——I
Hatching
I————I
Nestlings
I——I
Fledglings

Figure A1.4 Approximate timing of breeding for Whistling Kites in the high country.

Eggs: One to four eggs, mainly two, whitish or cream with fine brown flecks. Incubation lasts between 35 and 40 days.

Young: Nestling has cream first down, browner on the head where it is long and spiky and a pale fawn second down which is browner on the back. Young said to fledge in 44–54 days. The period of dependence lasts up to 2 months.

Voice and calls: Voice is distinctive: loud musical notes sliding down the scale, preceded or followed by rapid up-scale chatter.

Status: Not globally or nationally threatened. In the southern highlands populations show recent local declines, maybe because of drainage of wetlands and/or reduction in food supply, but possibly because of poisons, including Pindone, used to kill rabbits. Was nominated as a Threatened Species in the ACT, but the nomination was rejected.

Myths and questions: Often mistaken for other brown species of medium-sized raptor, such as Brown Falcons or juvenile Brown Goshawks but especially Little Eagles, which may have contaminated the observation data in the ACT.

Sources: Debus (2012); Fuentes *et al.* (2005); Marchant and Higgins (1993); Olsen (1992); Olsen *et al.* (2013d).

(Julian Robinson)

White-bellied Sea-Eagle

Haliaeetus leucogaster

Haliaeetus is from the Greek *haliaetos* a 'sea-eagle'; *leucogaster* is from the Greek for white-bellied. Raptors in the genus *Haliaeetus*, sea-eagles, are characterised by powerful feet and bare tarsi and include the North American Bald Eagle, Eurasian White-tailed Sea-Eagle, African Fish Eagle and the Siberian Steller's Sea-Eagle.

Field recognition

Signature characters: Unmistakeable large soaring eagle as big as a pelican or Wedge-tailed Eagle, pure white below, steel-grey back, loud honking call, broad upswept wings with black flight feathers, very short wedge-shaped tail; juveniles (up to 5 years) browner. Beak blue-grey at base, tipped with black, iris dark-brown, legs and feet cream. The slightly larger Wedge-tailed Eagle has longer primary feathers and a longer tail and feathered tarsi. **Adult** sea-eagles are grey, black and white, and are unmistakeable. **Juveniles** are brownish. At intermediate stages some individuals have a white head and underparts with a mottled brown breast-band like an Osprey. The White-bellied Sea-Eagle is 75–85 cm long, wingspan is 180–218 cm. Average weight of males is around 3200 g, females around 3800 g, ~90% as heavy as a Wedge-tailed Eagle.

Flight is a powerful stiff-winged action; they glide and soar with wings in strong dihedral or slightly kinked wings with curved trailing edges, showing white tail-end, except in first year birds. Perched they stand upright, angular-shaped, with broad shoulders narrowing to tail; wingtips extend to the end of or slightly past the tail-tip. Bare tarsi show when perched and in flight.

Distribution in the high country: Rare in the high country, four breeding pairs on Burrinjuck Dam, also Lakes Eucumbene and Talbingo in New South Wales. No confirmed breeding reported inside the ACT, except at Jervis Bay, though non-breeding foraging individuals and pairs ('floaters') are seen regularly along the Molonglo River, Murrumbidgee River, and on Cotter, Corin and Bendora Dams.

Habitat: Prefer coasts, islands, estuaries, rivers, dams and lakes. Sometimes they overfly forest and urban habitats.

Territoriality and behaviour: White-bellied Sea-Eagles aggressively defend their nest site against conspecifics and other predators. They perch conspicuously, perhaps as territorial display and make honking calls alone, in duets and in flight. In the breeding season pairs engage in mutual soaring, tandem flying and mock attack.

Food and hunting: They eat birds, reptiles, fish, mammals, crustaceans and carrion (Figure A1.5). In and around the ACT waterbirds and fish dominate their diet. They hunt these mainly from an exposed perch or by flying low over the water, followed by a quick dive onto prey. Unlike Ospreys, they seldom completely submerge when hunting fish, turtles, sea snakes and eels, instead scooping prey from near the surface. They also rob other predators.

Breeding: Across Australia the breeding season is broadly June to December. Local pairs usually start around August (Figure A1.6). In the high country they only breed near water, particularly along riverbanks and dams, in single pairs, not groups. The nest is a large pile of sticks 1–2 m wide, 0.5–1.8 m deep, unmistakeable except for a Wedge-tailed Eagle nest – the two eagles sometimes use the same nest

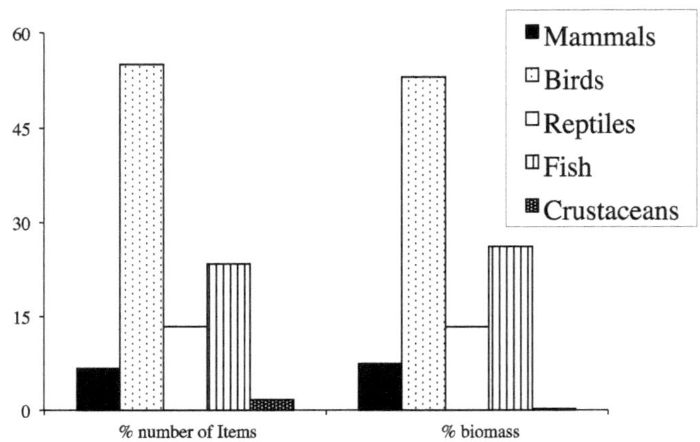

Figure A1.5 Diet of White-bellied Sea-Eagles in the ACT region (after Olsen *et al.* 2006b,c).

Mar April May June July Aug Sept Oct Nov Dec Jan Feb

I——I
Pre-breeding
I——I
Eggs
I——I
Hatching
I————I
Nestlings
I——I
Fledglings

Figure A1.6 Approximate timing of breeding for White-bellied Sea-Eagles in the high country.

in alternate years – lined with leaves, grass and, on the coast, seaweed, and often placed high in a large eucalypt or casuarina, 10–40 m above the ground.

We checked eight water bodies and their surface areas (dams when full) for breeding White-bellied Sea-Eagles and Whistling Kites between 2002 and 2010 in the high country near Canberra. Both human disturbance and prey availability appeared to affect their breeding, as well as the presence of Wedge-tailed Eagles that dominate them.

Eggs: One to two eggs, rarely three, dull white. Incubation takes ~40–42 days.

Young: The nestling has white down (see also first year above). The nestling period is said to be ~81–84 days.

Period of fledgling dependence lasts 2 or 3 months, perhaps longer, but this is largely unknown and needs further study. Young are said to leave their natal territory ~4–6 months after they fledge, and can disperse up to 3000 km. Age at first breeding is 7 years in captivity. No pairs have been found breeding in the wild in subadult plumage (before ~5 years old), although occasional individuals pair up or commence breeding with traces of immature plumage remaining (Debus 2012).

The one pair at Googong Dam, four in the Burrinjuck Dam area and one on the Murrumbidgee River near the west ACT border all show high reproductive success.

Voice and calls: Distinctive, deep goose-like honking or cackling; begging juveniles give more prolonged yelping or wailing.

Status: White-bellied Sea-Eagles are not globally or nationally threatened. In appropriate habitat, they can be relatively common for such a large predator though there are said to be local declines in southern Australia caused by habitat destruction and/or nest site disturbance. Shooting and other forms of direct persecution can be a local problem, but not in the ACT. However, the species is listed as Rare in the ACT.

Myths and questions: Although called an eagle, the species is closely related to the large kites of aquatic habitats, not to the true (booted) eagles. How much the declines in other parts of Australia are linked to lack of prey versus human disturbance or conflict with Wedge-tailed Eagles needs to be investigated.

Sources: Debus (2005, 2008, 2012); Olsen *et al.* (2006a); Dennis *et al.* (2011); Fuentes and Olsen unpublished data; Marchant and Higgins (1993); Olsen (1992); Olsen *et al.* (2013d); Olsen and Rehwinkel (1995); Corbett and Hertog (2011).

(Julian Robinson)

Swamp Harrier

Circus approximans

Circus is from the Greek *kirkos* for 'a circle' suggesting a kind of hawk or falcon that wheels or flies in circles. The common name 'harrier' is from the Old English *hergian* and refers to the chasing and harrying of prey. The specific name *approximans* is Latin for 'resembling' or 'approaching' (other harriers) perhaps the Northern Marsh Harrier specifically (Debus 2012).

Field recognition

Signature characters: Long, slim-bodied brownish raptor, larger than a raven, long wings, white rump, Barn Owl-like facial ruff, neck looks fatter than head, the only breeding harrier in the high country, though Spotted Harriers appear as vagrants.

Swamp Harriers are about the size of the Little Eagle or Whistling Kite, but distinguished from these species by their habit of sailing low over grassland, reeds and swamps on deeply dihedral wings, and turning or rocking and showing a prominent white rump patch. Swamp Harriers fly and glide with the wingtips widely fingered and turned up. Flight is buoyant, with smooth, rhythmic wing beats, long glides, and a slight rocking or tilting from side to side. They soar in circles and sometimes lower their legs in flight. They commonly perch on the ground or on low posts, stumps or swamp vegetation. At rest the wingtips reach almost to the tip of the long tail. Length is ~50–62 cm, wingspan ~120–145 cm. Average weight of adult males at Werribee, Victoria, was ~647 g, adult females 890 g.

Bill is black with a grey base, iris pale yellow, legs and feet yellow to orange but lemon-yellow in the non-breeding season.

Females are darker and browner. Older males are lighter, females greyer.

Adult Swamp Harriers can be confused with juvenile Spotted Harriers, which stray into the high country, but Spotted Harriers have a more ginger tone, lighter build, pale forewing patch, fawn (not white) rump patch, and a heavily barred tail with a wedge-shaped tip; they are more buoyant in flight, and adults have rufous barring on the front.

Perched, both harriers have long legs, whereas the large kites such as Black and Whistling Kites look 'legless' because of their short tarsi. Look for a light sandy head on Whistling Kites and charcoal on Black Kites compared with the brown-headed Swamp Harrier. Swamp Harriers have a squarish tail-end whereas

Whistling Kites are round-tailed, Little Eagles short square-tailed, Black Kites slightly fork-tailed. The Brown Falcon is smaller, with a prominent double cheek-mark, except for very dark juvenile falcons, and they lack the Swamp Harrier's pale rump patch.

Habitat: Open grassland, reed beds, heath and swampland. In the ACT they are now limited to Namadgi, especially Gudgenby and the Orroral Valley. They also breed near Hoskinstown and Collector. This species is a summer breeding migrant.

Territoriality: Very territorial; they defend a large area against other harriers and other raptors through displays, calling and chasing. Males mark their territorial boundaries by flying along them with the wings held high and the legs down. They show strong site fidelity.

Behaviour: The Swamp Harrier is solitary or loosely gregarious, usually seen quartering marshy or open country a few metres off the ground, often settling and perching on the ground. In the pre-breeding season pairs soar high and perform undulating displays with deep wing beats while twisting and giving *mew* calls. Males pass food to females in midair. The male occasionally dives at the female, which evades him or rolls to present her feet. Pairs sometimes lock talons. Females descend to the nest, males follow, descending in stages like a falling leaf. Wintering birds gather at communal roosts mostly outside the high country, and these roosts can serve as 'information centres' where some individuals follow others back to successful hunting areas.

Food: Swamp Harriers take mammals, birds, eggs, reptiles, amphibians, fish, insects and carrion (Figure A1.7). In the Canberra region, rabbits and birds dominate the diet. To capture prey they soar, quarter low over grassland and swampland then drop to the ground with legs extended and pin prey to the ground or snatch it and fly off. They rob other raptors.

Breeding: Breeding pairs were more common in and near Canberra 30 years ago, but only two nests have been found in the past 10 years, both in Namadgi.

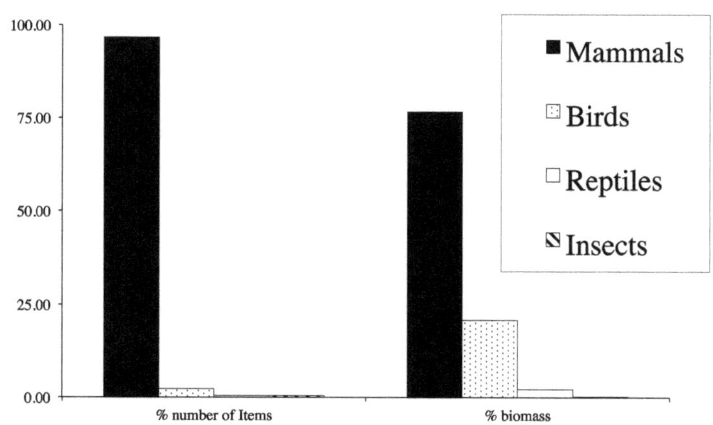

Figure A1.7 Swamp Harrier diet in the ACT (Olsen unpublished).

Mar April May June July Aug Sept Oct Nov Dec Jan Feb

```
                I——I
                Pre-breeding
                     I——I
                     Eggs
                          I——I
                          Hatching
                          I————I
                          Nestlings
                                    I——I
                                    Fledglings
```

Figure A1.8 Approximate timing of breeding for Swamp Harriers in the high country.

Laying season is September to December (Figure A1.8). Pairs nest solitarily or in loose 'clusters' with other pairs; occasionally polygynous. The nest is a platform of reeds, grass, sticks and other plants, and situated in tall grass, shrubs or reeds, either on the ground or in water.

Eggs: The clutch size is usually three or four eggs, ranging from two to seven, dull white. Incubation takes 33 days. In the high country they lay in September to November, and hatch young in October to December.

Young: The nestling has fawn to white first down with a white 'skullcap' and facial mask, and grey to buffy white second down. Young fledge 43 to 46 days later and hunt with the adults close to the nest before dispersing, about 4–6 weeks after fledging. Some juveniles wander or migrate up to 1500 km. Success has been measured as 1.05–1.75 young fledged per nest and 1.82–2.15 per successful nest.

Voice and calls: Soaring birds sometimes utter a high-pitched, descending *mew*.

Status: Not globally or nationally threatened, common in suitable habitat, but they have declined where wetlands have been drained. Nests disturbed by humans are often abandoned. There has been an appreciable decline in the ACT and the species should probably be listed as Vulnerable.

Myths and questions: Reasons for their decline as a breeding species in the ACT are unclear.

Sources: Baker-Gabb (1982, 1984); Debus (2012); Fuentes and Olsen unpublished data; Marchant and Higgins (1993); Olsen (1992).

(David Pastern)

Grey Goshawk

Accipiter novaehollandiae

Accipiter is Latin for a kind of hawk, one meaning is 'sparrowhawk'; *novaehollandiae* comes from 'Nova Hollandia' meaning New South Wales, denoting the country of origin of the first specimen known to science.

Field recognition

Signature characters: Striking grey-and-white or all-white raven-sized goshawk with bright orange-yellow cere and feet as **adults**. Wings are broader and more rounded than in the Brown Goshawk and legs and feet massive. Short wings and a long square-tipped tail. Iris deep red and legs and feet bright orange-yellow. **Juveniles**' upperparts darker, more brownish-grey with coarser wavy bars on the breast and a slight brown collar; eyes orange or brown.

The Grey Goshawk occurs in tall wet forest habitat in eastern and south-eastern Australia and riverine forest in northern Australia. Occasionally they appear in open woodland and urban areas, such as Canberra, but they are not known to breed in the ACT or in the high country. Their flight action is snappy but usually they appear slower than a Brown Goshawk, more owl-like than other raptors. Grey Goshawks soar and glide on short, bowed wings, not unlike a Sulphur-crested Cockatoo which they superficially resemble. The tail can be fanned. At rest they stand upright on long legs, their long tail extending well beyond their wingtips. The Grey Goshawk is 38–55 cm, the tail is slightly less than half the total length; wingspan is 71–110 cm. Average weight for males is ~350 g and females ~680 g (Tasmanian birds heavier: males ~415 g, females ~797 g). They are slightly larger and more heavily built than the Brown Goshawk with broader wings and shorter, squarer tails. Grey or white morphs occur in both sexes; only white morph occurs in Tasmania and it predominates in Victoria and western Australia, but both morphs occur as apparent vagrants in the high country.

Grey-morph Grey Goshawks are commonly mistaken for the much more common Brown Goshawk but Brown Goshawks have longer, narrower wings, a longer more rounded tail and rufous-coloured underparts in adults and striated breast in juveniles. *Elanus* kites can be mistaken for white-morph Grey Goshawks but have black marks on their underwings, longer more pointed wings, very short legs and they hover in flight, which goshawks do not.

Habits: In the breeding season Grey Goshawks perform undulating displays and pairs fly together using slow-flapping flights similar to those of the Brown Goshawk. In the non-breeding season, individuals can be seen flying and perching, chasing birds, boldly attempting to take domestic chickens or other birds around urban or rural houses.

Habitat: Both morphs of the Grey Goshawk are cryptic and difficult to see when perched high in rainforest with leaves shining white, or against giant pale-trunked forest gums. They often vocalise, betraying their presence. They perch on exposed limbs in tall trees on creek lines or at the forest edges, and are more readily seen when they soar in spirals above the forest.

Food: Prey ranges up to rabbit and heron size in southern Australia, including mammals such as rabbits and bandicoots, and birds, reptiles, amphibians, arthropods and (rarely) carrion. They forage mainly by still-hunting from a concealed perch in the tree canopy and by contour flying. They seize prey on the ground or a perch by diving, glides or direct flying attacks that become a short chase. They drag ringtail possums from their dreys (Nick Mooney pers. comm.). Diet in the high country is unknown.

Breeding: Laying season is September to December in southern Australia and March to November in the north. Pairs are solitary and defend territories. The nest is a platform of sticks 50–60 cm across, 35 cm deep, lined with green leaves, and placed up to 15 m above the ground in the canopy of a living tree. Timing of breeding in the high country is unknown. Longevity in captivity is up to 21 years.

Eggs: Clutch size is usually two or three eggs but occasionally four. Incubation takes ~30 days.

Young: The chick has white first down. Nestling period is 30–38 days. In one study, nine young fledged from 14 eggs; 94% of attempts were successful, with 1.5 young per successful nest and 1.4 young per active nest. The period of dependence after fledging lasts up to 6 weeks, after which the young disperse widely (Debus 2012).

Voice and calls: Though cryptic, they often vocalise, betraying their presence. Variety of mellow whistling notes and chattering calls given at rest and in flight. One call described as *weep-weep* or *swee swee*; also rapid *cac-cac-cac.*

Status conservation: The Grey Goshawk is not globally or nationally threatened and is common in the tropics and subtropics, but uncommon at the extremities of its range in southern and north-western Australia, particularly in the high country where it may be a rare or non-breeder. Habitat loss from farming and logging in south-eastern Australia may adversely affect them.

Sources: Debus (2012); Olsen unpublished data; Marchant and Higgins (1993); Olsen (1992); Olsen *et al.* (1990); Burton *et al.* (1994); Riddell (2011a).

Brown Goshawk

Accipiter fasciatus

Accipiter is Latin for a kind of hawk, one meaning is 'sparrowhawk'; *fasciatus* comes from *fascia*, a kind of bandage, swathe or bundle, and so 'banded' or 'striped', referring to barring on the breast.

Field recognition

Signature characters: Raven-sized raptor, **Adults** grey on the back, rufous barred on the front, with long round-tipped tail, slender, lanky-looking when perched, rufous breast, shy, usually concealed in foliage, heavy brow ridges give them a fierce look; **Juveniles** brown. Aggressive, active hawk that prefers to hunt and live in cover. They have relatively shorter middle toes than do sparrowhawks, and the tip of the middle toe does not project beyond the claws of the other toes. Iris pale to bright yellow and the legs and feet pale yellow.

The name 'Brown Goshawk' for a grey and rufous bird is a misnomer that causes many cases of misidentification. They are commonly mistaken for the much rarer Grey Goshawk that has shorter wings and tail and no rufous breast. Brown Goshawks and Collared Sparrowhawks hunt and nest in ACT woodland, and both are seen in the city, but Brown Goshawks are larger with rounded, not square, tails, proportionally longer wings, and they are more common than Collared Sparrowhawks. Brown Goshawk flight is rapid with bursts of quick wing beats; it glides with slightly bowed wings and soars. At rest it stands vertically on long legs, with its short wings not reaching the tail-tip. In contrast, a perched similar-sized Brown Falcon has a dark double malar stripe on its face, dark (not yellow) eyes, and long wings reaching to the tail-tip.

Brown Goshawks in southern Australia (race *fasciatus*) are 40–55 cm long (tail about half of this length) with a wingspan of 74–95 cm. The average weight of males is 350 g, females 570 g (Tasmanian females are heavier, ~614 g). Tropical birds (race *didimus*) are smaller: 37–46 cm long, with a wingspan of 70–85 cm and males weigh ~227 g, females ~343 g.

Juveniles especially range out onto grassland and through cities and towns, chasing birds and sometimes visiting aviaries.

Habitat: Brown Goshawks are secretive hawks of most wooded habitats, including mountains, farmland and trees in urban areas. In the high country they use a variety of habitats, but tend to nest in open woodland or, more commonly, in Red Stringybark open forest. Consequently they are fairly common (for a raptor) in Canberra Nature Park. They prefer eastern-facing hillsides for nesting, though they also use large eucalypts on flat ground, and it is not clear how high in the mountains they nest, though breeding pairs are seen to at least 1000 m.

Habits: They are wary and inconspicuous, often hiding behind clumps of leaves or tree trunks, or you glimpse one flashing through trees; easier to spot when soaring in spirals or calling loudly. In this they are very different from Collared Sparrowhawks which often ignore humans.

In cities or on farms they sometimes attack domestic birds including chickens. In the breeding season they perch more conspicuously, call, and soar together and perform flight displays, chasing or following each other with deep, exaggerated wing beats and calling with ringing whistles. Some defend their nests fiercely against intruders and will strike humans, others quietly sneak away and nests can be difficult to locate, even close to human settlement.

Around Canberra, Brown Goshawks disappear from their breeding territories in winter and may be partly migratory, though more study is needed. Some adults

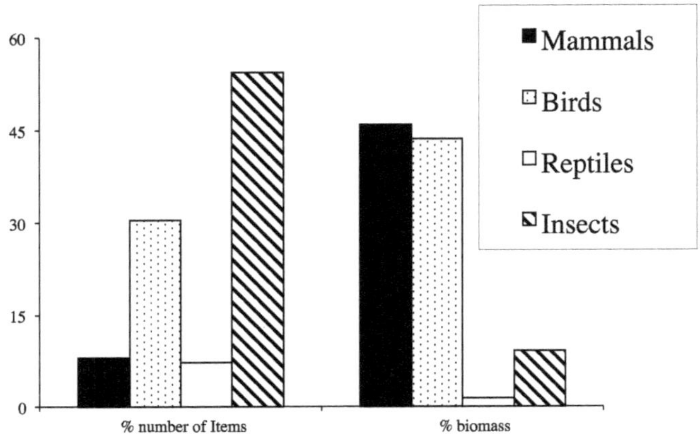

Figure A1.9 Diet of the Brown Goshawk in ACT region (Fuentes, Olsen and Rose unpublished).

are said to defend regular winter territories separate from their breeding territories, but this needs confirmation. From early autumn, a noticeable pulse of mostly young birds arrives in the cities and towns, maybe from local nests, and/or higher in the mountains.

Food: Brown Goshawks hunt mostly birds but also many small or medium-sized mammals, reptiles, amphibians and arthropods, and occasionally eat carrion. Around Canberra, they hunt rosellas, starlings, other birds, young rabbits, reptiles and large insects in woodland (Figure A1.9). They still-hunt from a concealed perch in foliage, soar or use fast contour-flying through forest using cover to ambush prey. They prospect by flying from point to point and can chase prey into bushes and snatch it. Prey usually weighs less than 500 g, but can weigh more than a kilogram, usually seized on the ground, from bushes or in midair. They take many invertebrates, in midair, arboreally, and on the ground.

Breeding: Laying season is September to December (Figure A1.10). Adults begin with slow-flying displays, often with light undertail coverts spread from the base of the tail, chattering and following each other. Around the ACT Brown Goshawks usually lay in a nest the size of a raven's nest, 38–70 cm across, 18–30 cm deep, lined with green leaves and placed 2–36 m above the ground in the fork of a living or, less often, dead tree in September or October. Young often fledge in December. They can nest 0.7 km apart in good habitat, for example, along the Molonglo River, but can be displaced by the larger, more powerful Little Eagle. Surprisingly, they can nest close to Wedge-tailed Eagles which may assist breeding Brown Goshawks by displacing Little Eagles. Sexual maturity is attained at one year, but breeding in juvenile plumage is rare. Mean annual survival is said to be 65–79% in south-eastern Australia, with banded birds living up to 17 years.

Mar April May June July Aug Sept Oct Nov Dec Jan Feb

I——I
Pre-breeding
I——I
Eggs
I——I
Hatching
I————I
Nestlings
I——I
Fledglings

Figure A1.10 Approximate timing of breeding for Brown Goshawks in the ACT region.

Eggs: Two to four eggs (usually three), dull white with brownish markings, incubated for ~30 days. Usually hatch during October or November in the high country.

Young: The chick has white first down and light salmon-brown second down with a white skullcap. The nestling period is 28–37 days. Young fledge at ~30 days and become very noisy, flying after parents screaming for food and chasing each other. Success was 1.6–1.75 young per attempt and 2.1–2.75 per successful nest in south-eastern Australia; 62% of attempts were successful in north-eastern Australia, with 1.6 young per successful nest and 1.0 young per active nest. The period of dependence after fledging lasts up to 6 weeks, then juveniles disperse or migrate up to 900 km (Burton *et al.* 1994; Debus 2012). After independence the young commonly harass birds in aviaries or chicken runs or pigeon lofts, much to the annoyance of owners.

Voice and calls: A variety of rapid chattering *cac-cac-cac* calls when attacking enemies but a *swee* call when courting or when males deliver food to females. Females, and perhaps large young, give a sharp one-note call not unlike the call of a Sacred Kingfisher, repeated about one note per second. This call can help researchers find nests.

Status conservation: Brown Goshawks are not globally or nationally threatened. For a raptor they are common in the ACT region, but in recent years, especially after the 2003 fire, they have declined. There have other been local declines in southern Australia where habitat clearance is extensive, but probably Brown Goshawks have benefited from the introduction of European Rabbits, starlings and perhaps Common Mynas. People shoot them (mainly juveniles) but this is rare in the high country and would not affect overall populations.

Sources: Debus (2012); Fuentes and Olsen unpublished data; Marchant and Higgins (1993); Olsen (1992); Burton *et al.* (1994); Aumann (1988, 1989, 2001); Riddell (2011b).

Collared Sparrowhawk

Accipiter cirrocephalus

Accipiter is Latin for a kind of 'hawk' or 'sparrowhawk'. The specific name *cirrocephalus* combines the Greek for yellow or tawny and headed referring to the rufous half-collar around the hind neck.

Field recognition

Signature characters: Like a miniature Brown Goshawk, a tame little hawk with staring yellow eyes. **Adults**: grey, rufous front, female is about the size of a Feral Pigeon with longer, square-end tail and shorter, broader wings, male about the size of a Magpie-lark or Nankeen Kestrel, tame, chases birds in cities and woodland; **juveniles** browner. It has a square tail-end, slightly forked when at rest, and is less common than Brown Goshawks in the high country. The Collared Sparrowhawk has slight brow ridges and long slender legs. The middle toe projects beyond the claws of the other toes. Iris pale yellow, the legs and feet pale yellow, ~30–40 cm long, their tail makes up about half of this, with a wingspan of 55–80 cm. The average male weighs ~125 g, females ~240 g. The female is similar in size to a male Brown Goshawk but slimmer.

Collared Sparrowhawks fly with quick wing beats in low direct flight often from tree to tree. They glide on flat or slightly drooped wingtips, and soar with their wings outstretched and intermittent shallow wing flickering. Perched they stand upright on long legs; wingtips don't reach the tip of the tail.

In flight it is best distinguished from the Brown Goshawk by its smaller head, and square or notched tail-tip. At rest the sparrowhawk has more staring eyes (it lacks the Brown Goshawk's heavy brow ridges), more spindly legs, and relatively longer middle toes. The similar-sized Australian Hobby, a fast-flying falcon, usually flies above the woodland canopy, not below it, has a dark head pattern, dark eyes, long, narrow pointed wings (reaching the tail-tip at rest) and shorter legs.

Habitat: Wooded habitats and forests, including farmland, city parks, mallee woodland and creek lines in the inland. Regionally they prefer open woodland and parklands particularly introduced pines. They also use urban areas, sports ovals and golf courses. They also breed at higher elevations in the high country.

Behaviour: The Collared Sparrowhawk can be tame and approachable when perched, though it is cryptic and easily overlooked. Mobbing birds such as friarbirds, Noisy Miners and currawongs can betray their presence, as does the sparrowhawk's quick, chattering *cac-cac-cac* calling when mobbed, or when driving enemies from the nest area. Early in the breeding season, individuals or pairs soar, perform displays, give chittering calls, and perch in the nesting area and give slower, mellow *mew* calls. Some food transfers are aerial.

After the breeding season, some adults and juveniles move into urban areas where they may hang around aviaries and attack birds at bird feeders. They are less prone to conflict with humans than are Brown Goshawks, which take poultry.

Territoriality and behaviour: Sparrowhawks noisily defend their eggs, young and prey from currawongs and ravens, and defend territories from other sparrowhawks.

Food: Collared Sparrowhawks are expert ambush hunters. They hunt by flying to a perch and waiting in a concealed position in foliage, then making short tree-to-tree flights to other concealed positions. They also fly low and fast, trying to surprise prey. They seize prey in flight after a direct flying attack or after a direct glide that becomes a short chase. They also attack prey on the ground. Most prey weighs less than 100 g, rarely more than 200 g. They hunt mainly small birds weighing 20–75 g, especially passerines, including introduced sparrows, mynas and starlings, and insects, rarely lizards and small mammals. Around Canberra the main prey items are finches and other small passerines, especially House Sparrows, as well as insects, especially beetles (Figure A1.11). They have a high metabolic rate so need plenty of food and they have a proportionally large crop that makes them look top-heavy when the crop is full.

Breeding: The breeding season in Australia is July to February (Figure A1.12). They build a magpie-sized nest 27–32 cm across, 12–15 cm deep, lined with green leaves and placed 4–39 m above the ground in the fork of a living eucalypt or, more often, an introduced pine tree, including unobtrusively in well-wooded urban parks. This hawk, and the Australian Hobby, breeds later in the season than do

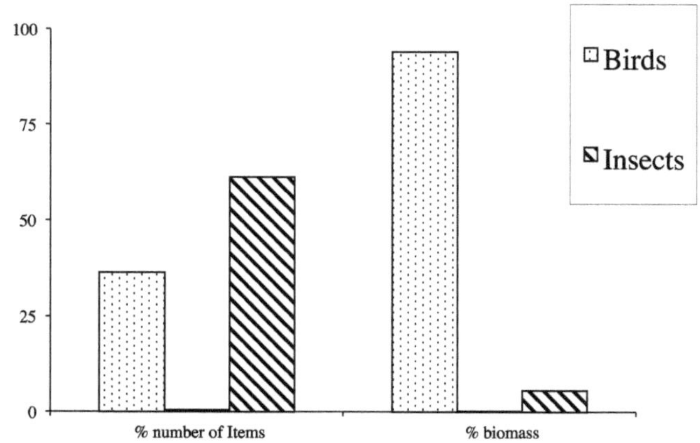

Figure A1.11 Diet of the Collared Sparrowhawk in the ACT (Fuentes, Olsen and Rose unpublished).

Mar April May June July Aug Sept Oct Nov Dec Jan Feb Mar

I——I
Pre-breeding
I——I
Eggs
I——I
Hatching
I————I
Nestlings
I——I
Fledglings

Figure A1.12 Approximate timing of breeding for Collared Sparrowhawks in the ACT.

other raptors in the high country. It tolerates human disturbance during breeding if the nest is high enough in the tree.

Eggs: In the high country three or four eggs, usually three, dull white, sometimes finely speckled with brown, are laid in November. Incubation takes ~35 days.

Young: Young fledge in ~28–33 days in January. The chick has a white first down and greyer second down with a white 'skullcap'. Success in one study was 1.6 young fledged per clutch laid, and 1.7 per successful nest; in central Australia success was 2.1–3.25 young per active nest, with failure in a dry year. Nestlings are occasionally taken by other raptors. The period of dependence after fledging lasts up to 6 weeks, after which young disperse. Sexual maturity is attained at one year, with birds sometimes breeding in juvenile plumage. The species has a low reproductive success around Canberra, perhaps because of interference from currawongs and/or ravens.

Voice and calls: Calls are more rapid, thinner and higher pitched than those of the Brown Goshawk. Its most commonly heard call is a very rapid chitter not unlike the alarm call of the White-plumed Honeyeater *Lichenostomus penicillatus*; also a series of single notes not unlike the Sacred Kingfisher but longer, sometimes breaking into a chitter. Begging juveniles give a distinctive loud, slow *mew* repeated in a series, about one note per second and chittering calls as they chase others in play. The male gives soft mewing notes when he brings food to the female which seizes the food and gives loud single-note 'dismissal' calls similar to the begging calls of fledglings.

Status: The Collared Sparrowhawk is not considered globally or nationally threatened. Although widespread, it is uncommon over most of its range, though probably overlooked. In the south-eastern highlands and the Canberra region, its reproductive success may be affected by increases in the populations of Pied Currawongs and ravens, competitors that can rob and injure adults and kill nestlings. Sparrowhawks often collide with windows and are brought into rehabilitation centres.

Sources: Debus (2012); Fuentes and Olsen unpublished data; Marchant and Higgins (1993); Olsen (1992); Debus *et al.* (1993); Aumann (2001).

Wedge-tailed Eagle

Aquila audax

Aquila is Latin for 'eagle', *audax* is for 'bold' i.e. 'bold eagle', either a misnomer for this normally shy bird, or a reference to its hunting behaviour.

Field recognition

Signature characters: The largest raptor in Australia, a huge dark eagle with long upswept wings, fingered wingtips, long boat-shaped or diamond-shaped tail, considerably larger than a raven, Little Eagle or Whistling Kite. Baggy feathered legs when perched or running on the ground, lanky build with a large bill, small head and long neck.

Iris brown, feet creamy-white. Older birds (5 to 6 years old) are darker, have a reddish-brown nape and back and much narrower reddish-brown upperwing bar.

They are ~85–104 cm long, their tail makes up about two-fifths of the body length, their wingspan is ~186–227 cm. Average males weigh ~2735 g, females ~3950 g. Perched, its shoulders look broad, giving it a powerful look. It soars and glides with wings held in a dihedral. Even in gale-force winds the flight of this eagle is fairly stable and controlled but it can rock a little from side to side when turning.

Wedge-tailed Eagles in flight can be confused with juvenile White-bellied Sea-Eagles that have shorter, broader, more stiffly upswept wings, a short pale tail and bare tarsi.

The Wedge-tailed Eagle can be solitary but juveniles especially can gather at road kills or join other eagles in flight; two or more eagles can join together to attack large macropods.

Habitat: Most terrestrial habitats except intensively settled or cultivated areas. In the high country they often nest on steep hillsides (mainly facing north-east) and gullies.

Territoriality and behaviour: Wedge-tailed Eagles perch prominently in tall, often dead trees, on rocks or cliff-tops, or shrubs or low trees, but usually seen soaring high on thermals, typically above hill crests or escarpments, but also over flat plains. They are very territorial and defend a large area against other Wedge-tailed Eagles with prominent perching, high soaring and 'boundary-riding' or 'pot-hook' displays, also called 'sky dancing'. In these undulating displays the eagle folds its wings, swoops down, then at the bottom or pot-hook shoots upwards and, at an apogee, back down like a roller coaster. They can stoop at an intruding eagle, causing one or both to roll and present its claws. They attack other raptors such as Little Eagles or Brown Falcons, perhaps as part of territoriality, but maybe as food.

They sometimes attack hang-gliders near their nests, especially near the Lake George escarpment, where they rip fabric of these aircraft and endanger pilots, though some of these may be juveniles playing. Pilots have tried to use air-horns and other devices, but to little effect. Wedge-tails show strong site fidelity, and territories in the ACT region have been occupied for at least 40 years, probably much longer, and probably by successive birds.

Breeding pairs perch together, extend their necks, touch beaks and allopreen. The male may present nesting material, sometimes before copulation. At all times of year, but particularly in the breeding season, pairs spiral up together and can descend to mate on a branch near the nest.

They are often seen at road kills, though it is policy in the ACT to remove road-killed macropods from roadsides so eagles and other carnivores aren't struck by cars.

Food and hunting: Take mammals, birds, reptiles and carrion. Because Australia has no vultures, they fill the niche of large vultures elsewhere in the world. They are believed to prefer rabbits and hares, as substitutes for the same-sized macropods now extinct in the high country and elsewhere in much of Australia, but they probably evolved to kill mainly macropods and birds. Rabbits are important prey in arid Australia, as are road-killed kangaroos on highways.

Around 40 years ago in the Canberra region they took many rabbits (around 50% of the diet by number), but nowadays they eat mainly macropods and large birds (more than 70% by biomass, see Chapter 9) when both groups are combined. They commonly take crows, cockatoos, waterfowl, dragons and goannas, and sometimes snakes (Figure A1.13). Lambs are rarely killed, but sometimes eaten as carrion (as revealed by skinning a dead lamb to reveal the haemorrhage caused by eagle talons, or death caused by other predators such as dogs). They also take possums and pirate prey from other raptors.

They forage by slow prospecting, high soaring or still-hunting from a perch. Mostly they seize prey on the ground, less often from the tree canopy. They use a long, swift glide with wings partly closed and wrists forward, or stoop and then chase; they rarely take prey in flight. Some attacks on kangaroos appear to be cooperative. They gather at kangaroo or other large animal carcasses and dominate other scavengers such as kites or corvids.

Breeding: Laying season is April to September, earlier in the tropical north (Figure A1.14). The nest is a huge accumulation of dead sticks typically 70–90 cm across, 30–80 cm thick, but often larger than 1.8 m across and 3 m high (one on the Murrumbidgee River because of repeated use); said to weigh 400 kg but this could be conservative; inner cup is lined with green leaves. They usually nest in a live or dead tree with a commanding view, 2–73 m above the ground, in the tallest tree available. They don't seem to be able to build nests on cliffs, as the closely related Golden Eagle does in the Northern Hemisphere and Verreaux's Eagle does in

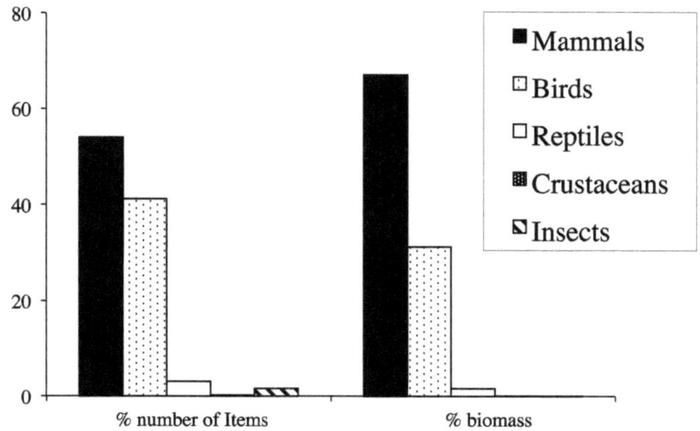

Figure A1.13 Diet of the Wedge-tailed Eagle in the Canberra region (Fuentes, Olsen and Rose unpublished).

Africa, but, on the South Australian coast, they occasionally use nests on cliff ledges originally built by White-bellied Sea-Eagles. They occasionally build on artificial structures such as power poles or pylons. Almost always the nest is below a ridge top, allowing them to glide into the nest with heavy items instead of 'carting' prey uphill. Although they usually build or repair several nests in an area, they tend to favour one per season, though another nest may be used as a feeding platform.

Eggs: Clutch size is usually two eggs, ranging from one to three, very rarely four, dull white freckled with purplish to reddish-brown; in Tasmania the clutch size is usually one. Near Canberra they tend to lay two eggs in June to August, incubate them for ~45 days, and hatch young in August or September. Incubation takes ~43 days per egg. Success varies regionally and with years, as low as 0.19 young per year in south-western Australia on a rabbit-dominated diet to ~1.20 young per pair per year in Queensland on a macropod-dominated diet, and ~1.1 young per pair per year in recent studies in the high country, also on a macropod-dominated diet.

Young: The chick has white down (including on the tarsi) which is long and hair-like on the head, and a prominent black bill. The nestling period is ~79–95

Mar April May June July Aug Sept Oct Nov Dec Jan Feb

I——I
Pre-breeding
I——I
Eggs
I——I
Hatching
I————I
Nestlings
I————I
Fledglings

Figure A1.14 Approximate timing of breeding for Wedge-tailed Eagles in the ACT.

days. In the highlands they fledge one or two young about 12 or 13 weeks after hatching from October to December. The period of dependence after fledging is said to last 3–6 months, but this needs study; young can disperse widely, commonly 200 km but up to 800 km. Groups gather in late summer on the Murrumbidgee River around Wee Jasper. Wedge-tailed Eagles can breed at 3 years old, and birds will pair in immature plumage and perch together, but they seldom attempt breeding before acquiring adult plumage. The oldest banded wild bird was 11 years old, the oldest in captivity ~40 years.

Voice and calls: Usually silent, but they utter sharp yelps and squeals when food-begging, courting, or against enemies.

Status: Not threatened globally or nationally, with the exception of the endangered Tasmanian subspecies, *Aquila audax fleayi,* which is classified as Endangered because it has been persecuted and its population is small, about 300 breeding pairs. Protecting nests with buffer zones and restrictions on disturbance during the breeding season can help, but some birds move to alternative nests up to 5 km away, so a buffer around the previous year's nest can be counterproductive. There have been no appreciable declines after the introduction of the Rabbit Haemorrhagic Disease Virus (RHDV) and subsequent decrease in rabbit numbers, though this virus has been less successful in the high country than in other parts of Australia. Wedge-tailed Eagles are still subject to illegal shooting and poisoning, even in the ACT region. There have been local declines in the south due to habitat destruction and human disturbance in heavily settled areas; they do not tolerate human activity and will abandon the nest. They may have benefited from thinning forest cover, introduction of the rabbit, increasing macropod numbers and domestic animal carrion.

Breeding pairs were common near Canberra 30 years ago, but many disappeared with the expansion of the city. Some pairs have returned to suburban areas in recent years, mainly because of the suitable farmland, availability of tall trees for nesting, increasing numbers of macropods, and foraging grounds provided by Canberra Nature Park. For a large predator, they remain common throughout the high country.

Myths and questions: Several myths persist about Wedge-tailed Eagles and rabbits, for example, that eagles need rabbits to breed, that eagle numbers correlate with rabbit numbers, that eagle breeding success correlates with rabbit numbers, that eagles prefer rabbits over other prey. None of these claims are supported by data. In Australia it is not clear why no other raptors besides Peregrines and kestrels nest on cliffs, while closely related eagle species in the Northern Hemisphere regularly do. Claims that males courtship feed females are not supported by field observations.

Sources: Debus (2012); Fuentes and Olsen unpublished data; Leopold and Wolfe (1970); Marchant and Higgins (1993); Olsen (1992); Olsen *et al.* (2010a,

2013a); Debus *et al.* (2007a); Fuentes *et al.* (2007); Cherriman *et al.* (2009); Foster and Wallis (2010).

(Julian Robinson)

Little Eagle

Hieraaetus morphnoides

The Latin *Hieraaetus* is from the Greek words *hierax* (hawk or falcon) and *aetos* (an eagle); *morphnoides* means 'eagle-like', in its diminutive form, implies a likeness to larger eagles.

Field recognition

Signature characters: Stocky, small soaring eagle a bit larger than a raven, broad wings, small crest on broad head when perched, feathered legs, dark upper body, lighter underparts, except in the dark morph.

Little Eagles come in two or three colour morphs, depending on treatment: light morphs are light cream underneath and dark brown above, with a striking underwing pattern; dark morphs are dark sooty-coloured all over with pale inner primaries. Rufous birds are much redder underneath with heavier streaking below but, if only two morphs exist, these may simply be aberrant light morphs. This polymorphism occurs in both sexes and can occur within a brood of juveniles.

The average weight of males is 635 g, females 1046 g, so they are highly dimorphic by weight, but this is less obvious in flight because males have proportionally longer wings compared to their body size. The sexes can thus be difficult to separate in flight unless a pair is together. Little Eagles tend to occur in pairs or alone, not in groups. In direct flight they are powerful and swift, and capable of stoops with closed wings. In soaring and gliding flight they show the characteristic, pale upperwing band and a rather short, square-tipped tail, which sometimes is fanned. They glide on flat wings with the primaries slightly dipped and soar with the wings held flat, sometimes with primaries slightly swept back. Little Eagles are ~44 to 55 cm long, the tail makes up roughly one-third of its length; wingspan is ~110 to 136 cm. Little Eagles are distinguished from the similar-sized Whistling Kite by being stockier, with shorter wings and tail, which is square-cut, a different underwing pattern, and a short, broad head and a stubby tail. The Whistling Kite has longer more separated primaries, and a jerkier, less shallow flight action; perched, Whistling Kites have bare tarsi and relatively small feet.

The Black Kite is more slender with longer primaries, longer forked tail, and smaller feet; the Swamp Harrier has longer wings, white rump patch, and tends to quarter close to the ground, which Little Eagles seldom do.

Habitat: Prime habitat is characterised by woodland, scrub in rough or hilly country, gum trees along creek lines in inland Australia, and riparian country up to ~1500 m above sea level but breeding probably only about up to 1000 m. Near Canberra they prefer river country and open woodland; they usually avoid dense forest and have not been found nesting at high elevations in Namadgi or Kosciuszko National Parks.

Territoriality and behaviour: They can be easily overlooked when perched. Little Eagles spend much time soaring high and their chirping call is often heard before the eagle is seen. They commonly attack other raptors by flying above them and diving at them, while calling.

Its display flight is a distinctive series of undulations producing an obvious flash pattern in pale birds, with the descending chirping call given loudly and frequently. Members of a pair soar together, sometimes culminating in a long dive to the tree canopy or nesting site. Food-begging females squeal and chitter loudly at the male.

Food and hunting: They hunt medium-sized mammals, birds, reptiles and large insects, and take some carrion and, rarely, fish. In southern Australia they take, by number and biomass, mostly young rabbits; in the north where rabbits are scarce or absent, they take mainly birds and arid-zone lizards, up to ~1.5 kg in weight. They forage by prospecting and soaring, by low quick flights between perches, and by still-hunting from a perch. They snatch prey from the ground from a long glide or dive, sometimes stooping into the tree canopy. They are capable of rapid twisting pursuit and stoops, and near Canberra they catch mainly rabbits, and small to medium-sized birds up to the size of ducks, especially parrots and corvids (Figure A1.15).

Breeding: Egg laying varies with latitude, earlier and longer in the tropical north often during the dry season (March to September), shorter in the dry centre

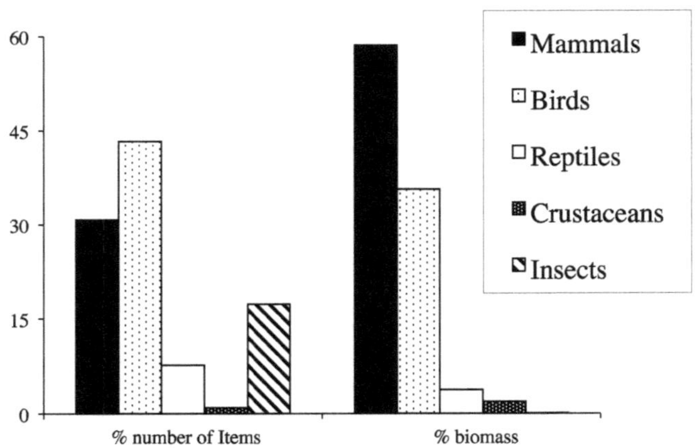

Figure A1.15 Food of Little Eagles in the high country (Fuentes, Olsen and Rose unpublished).

```
Mar April May June July Aug Sept Oct Nov Dec Jan Feb

                    I——I
                    Pre-breeding
                        I——I
                        Eggs
                            I——I
                            Hatching
                            I————I
                            Nestlings
                                    I——I
                                    Fledglings
```

Figure A1.16 Approximate timing of breeding for Little Eagles in the high country.

and south (usually August to October; rarely starting in May or extending to December for replacement clutches). In the high country Little Eagles lay eggs in late August–early September (Figure A1.16). The nest is as large as a raven's nest; outside diameters of most nests were ~60–75 cm, and ~30 cm deep, lined with green leaves, and placed 5–45 m above the ground in the fork of a living or dead tree. They sometimes use an abandoned Whistling Kite or Wedge-tailed Eagle's nest.

Eggs: The clutch size is usually two eggs, but it ranges from one to three, dull-white, sparingly spotted and blotched with pale reddish or purplish-brown. Incubation takes ~36–39 days.

Young: The nestling has white down (including on the tarsi) which is long and hair-like on the head and buff on the belly; it also has a dark 'mask' around the eyes. Though Little Eagles are close to the weight of Peregrines or Black Falcons, they have smaller clutch sizes, and longer incubation and nestling periods than large falcons because Little Eagles are taxonomically close to large eagles like Wedge-tails. They fledge one or two young (average brood size is 1.1 young) after ~52–66 days in the nest. Breeding success varies regionally and according to prey availability and weather, said to be lower in wet years or drought years, from 0.5 to 1.0 fledged young per pair per year; 0.8–0.9 young per clutch laid; 1.1 young per successful nest. The period of dependence after fledging lasts about 2 months, after which young disperse or migrate up to 2884 km (Debus pers. comm.). The oldest banded wild bird in Australia was banded at CSIRO in Gungahlin and found dead near its nest on Black Mountain 26 years later.

Voice and calls: Its most common call is distinctive: a loud, rapid excited two- or three-note descending chirp while soaring.

Status: The Little Eagle is said not to be globally or nationally threatened, but the species has not been carefully monitored. Local populations are declining, partly because of urban expansion and maybe because of an increase in Wedge-tailed Eagles, but maybe because of poisons such as Pindone. It is listed as

Vulnerable in the ACT and New South Wales (Olsen *et al.* 2013a). In 2011 we found only one breeding pair, so their status should probably be elevated to Endangered in the ACT (Olsen *et al.* 2013b). Some pairs are tame and not as easily disturbed at the nest as are White-bellied Sea-Eagles or Wedge-tailed Eagles, but certain pairs can be easily disturbed and human-caused nest failures do occur. They are seldom persecuted.

Myths and questions: This is one of the smallest eagles in the world. In the early 1990s it was common in the ACT region and their decline has corresponded with an increase in rabbits, so Pindone and other chemicals may be implicated. They are not as rabbit-dependent as has been claimed.

Sources: Debus (2012); Fuentes and Olsen unpublished data; Gjershaug *et al.* (2009); Marchant and Higgins (1993); Olsen (1992); Olsen *et al.* (2010a); Debus *et al.* (2007b, 2013); Debus and Ley (2009); Debus (2011); Olsen *et al.* (2013b).

(Geoff Dabb)

Nankeen Kestrel

Falco cenchroides

The specific name *cenchroides* probably alludes to this kestrel's similarity to the Eurasian Kestrel *Falco tinnunculus* of Eurasia and Africa. Nankeen Kestrels may form a super-group with other kestrel species such as Eurasian Kestrel and Spotted Kestrel *F. moluccensis*.

Field recognition

Signature characters: Slightly smaller than a Feral Pigeon with long narrower, wings and tail; back is chestnut or reddish-brick coloured, front is whitish cream, black wingtips, black band near the tail-tip.

These small, delicate falcons have moderately long tarsi and relatively short toes, are ~30 to 35 cm long, with the tail making up half of the total length; wingspan ~60 to 80 cm. The average weight of the males is 165 g, females 185 g. It is the smallest Australian raptor, other than the male Collared Sparrowhawk, but Hobbies can look just as small because they are about the same length and dark birds can look smaller than light ones.

Adult males have a grey head and tail, females are chestnut. Iris brown, legs and feet yellow. **Juveniles** similar to adult female.

The kestrel's wing action is quick and flickering, with glides on flat wings; it hovers or hangs in the wind with wings flexed (kiting), head steady to survey the terrain below. Black-shouldered Kite also hovers, but the kestrel is smaller and has more buff colour, less white than the Black-shouldered Kite, with narrower wings, no black underwing markings and longer, round-tipped tail with black sub-terminal band. Kestrels make angled stoops with closed wings onto ground-dwelling prey whereas Black-shouldered Kites parachute more vertically onto prey. Pale Brown Falcons at rest can be confused with kestrels but are larger with a double malar stripe, pale cheek-mark and longer legs. In poor lighting the kestrel can be confused with the Australian Hobby, which is darker with more pointed wings, a dark partial head helmet and a shorter, squarer tail.

Habitat: Often seen along roadsides, on fences, wires, power poles and dead trees, or hovering above paddocks, although it is also common in treeless inland areas away from roads. Characteristic of farmland with scattered trees and inland shrublands and city parks, they occur in most open habitats including semi-open woodland, but avoid heavy forest. Around cities and farms they commonly perch on buildings. They nest on Mt Kosciuszko, higher than any other Australian raptor.

Habits: Nankeen Kestrels defend their nests aggressively, swooping at and striking intruders. They perform a winnowing display with bursts of short, quick wing beats, gliding on drooped wings, while chattering. They sometimes twist the body from side to side in flight to produce a white flash pattern. Members of a pair soar together; on perches they bow to each other and use *chip-chip* calls and food-begging *trill* calls. Males pass food to females at perches near the nest or sometimes in midair.

Food and hunting: Nankeen Kestrels take many invertebrates, particularly insects such as grasshoppers and crickets but also spiders and other arthropods and small mammals such as mice, birds up to pipit, sparrow, starling and rosella size, and reptiles such as skinks and small dragons, especially when breeding. Near Canberra they eat mainly small dragon lizards, grasshoppers, Christmas beetles and other insects (Figure A1.17). Like other falcons, these kestrels tend to eat smaller prey like insects early in the breeding season then switch to larger prey such as birds when the young hatch. They commonly still-hunt from telephone wires, poles, treetops or forage by prospecting flights and hovering. They dive onto and seize prey on the ground in a shallow dive or sometimes in a vertical stoop and they occasionally tail-chase birds and catch them in midair. They also hawk insects in midair.

Breeding: The laying season in Australia is August to December (Figure A1.18). Pairs usually nest alone, but sometimes semi-colonially where nesting habitat is clumped or there is abundant prey, such as woodlots in farmland. They nest in sink-holes or mine shafts below ground but mostly in tree hollows, old nests of other birds, cliffs ledges and potholes, buildings or machinery, 2 to 31 m above ground. Near Canberra they usually nest in tree hollows, and occasionally in potholes on cliffs or on building ledges, and lay one to six eggs (average 3.85), usually in late October or November; incubation takes 28 days per egg, ~30 days for the clutch, with ~66% hatching success (of eggs laid).

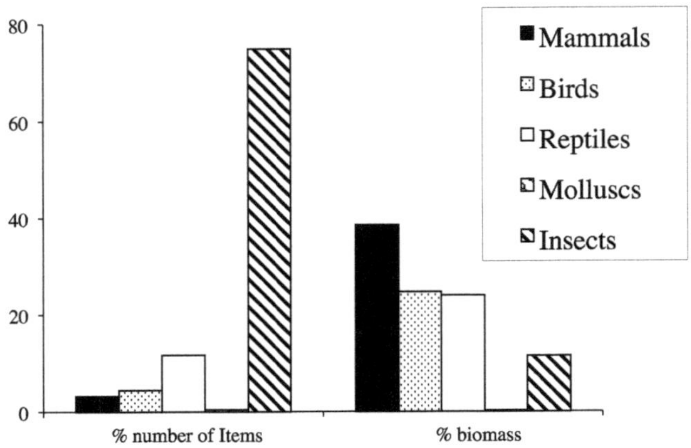

Figure A1.17 Diet of the Nankeen Kestrel around the ACT (Fuentes, Olsen and Rose unpublished).

Eggs: Eggs are pale white or buff-coloured, spotted and blotched with reddish to purplish-brown, laid in a depression or scrape in dirt on a cliff or wood and debris in a tree hollow. The clutch size is usually three to five eggs, ranging from one to six, average 4.3.

Young: Near Canberra the young fledge in December or January after ~31 to 35 days in the nest. Mostly they produce one brood a year, but in good seasons they can produce two. The chick has white down. Nestling period is 31 to 35 days. Fledging success has been measured as 68% of breeding attempts, 45% fledging success of eggs laid, and 1.26 young fledged per breeding attempt. The fledged young stay near the nest and are fed by parents for 11 to 12 weeks. The period of dependence after fledging lasts up to 2 months, after which the young disperse or migrate widely; up to 800 km has been recorded, though usually within 10 km of the banding site. Around Canberra they fledge on average 1.6 to 2.1 young per territory. Young are sexually mature and sometimes breed at a year old. Annual survival of adults is at least 60%.

Voice and calls: Its most commonly heard call is a rapid, high-pitched *cac-cac-cac* (three notes per second); also a *chip-chip-chip* repeated (one note per second) and food-begging *trill* whining from females and juveniles.

Mar April May June July Aug Sept Oct Nov Dec Jan Feb

I—I
Pre-breeding
I—I
Eggs
I—I
Hatching
I———I
Nestlings
I—I
Fledglings

Figure A1.18 Approximate timing of breeding of Nankeen Kestrels around the ACT.

Status: Not considered threatened globally or nationally. Abundant and widespread throughout its range, where it might have increased in numbers, perhaps benefiting from clearing and the introduction of mice, starlings and other prey. Kestrels have now settled on Lord Howe, Norfolk and Christmas Islands and maybe New Zealand.

Myths and questions: The only diurnal raptor that nests in hollows, except for Peregrines in certain parts of their range and, with Peregrines, the only inland diurnal raptors to nest on cliffs.

Sources: Bollen (1991); Debus (2012); Olsen and Fuentes unpublished data; Marchant and Higgins (1993); Olsen (1992); Olsen and Olsen (1980); Paull (1991); Baker *et al.* (1997); Aumann (2001); Starr *et al.* (2004).

(Julian Robinson)

Brown Falcon

Falco berigora

The species name *berigora* is an Aboriginal word for this falcon.

Field recognition

Signature characters: Raven-sized raptor with more rounded wingtips than most falcons. **Adults** vary in colour from dark brown to light cream but, unlike other falcons or other raptors they always have these four characteristics: 1) a double malar stripe on the side of their face enclosing a pale cheek patch; 2) red-buff barring and spots on their flight and tail feathers; 3) underwing flight feathers are pale, contrasting with darker underwing coverts; and 4) red-brown to dark 'trousers' on the thighs. Though many texts consider different colour forms to be different morphs or even subspecies, the different plumage colours in Brown Falcons may reflect different ages and sexes of the same morph. In many birds the underparts become whiter with age, especially with males. Eyes brown or (rarely) hazel, legs and feet pale blue-grey. **Juveniles** dark brown.

The Brown Falcon is ~40–50 cm long, the tail almost half of the total length, with a wingspan of ~90–120 cm. The average weight of males is ~470 g, females ~625 g. When perched, Brown Falcons appear more long-legged than other falcons, characterised by heavy scutellate plates on the front of the tarsi and reticulate scaling behind, said to protect them from poisonous snakes they prey on, and probably useful for running through long grass and other vegetation. They also have short toes suited to capturing other reptiles, and running and foraging on the ground.

They are similar in size and shape to a Brown Goshawk, with which they are often confused, but in addition to the four key characters above, Brown Falcons have slightly longer wings and a chicken-like cackling voice, which no other falcon or hawk has. The wingtips are not as pointed and knife-like as in other falcons, and the tail-tip is rounded, in contrast to the more squared tail-tip in Peregrines and Hobbies. Pale adult Brown Falcons can be confused with Nankeen Kestrels, but see the four characters above; Brown Falcons are also larger and browner with longer legs. Falcons as a group tend to be dashing and quick, but Brown Falcons can be sluggish instead of dashing, so early bird observers insisted on calling them Brown Hawks. Their flapping flight is slow and more laboured, similar to a Northern Hemisphere *Buteo* or buzzard species; Brown Falcons are their ecological equivalents though taxonomically distant from *Buteos*. Brown Falcons glide on wings raised in a dihedral. When rodent prey is available, they hover, rather heavily compared with the Black-shouldered Kite or Nankeen Kestrel, and much less often (no other falcon except the kestrel hovers); they are expert at kiting in a headwind. Perched they have a pot-bellied profile, rather loose-feathered with a large head, rounded shoulders (not square as in Black or Peregrine Falcons), and longer legs. The dark Brown Falcon is often mistaken for the Black Falcon which has more pointed wings, faster flight, a square-tipped tail, shorter legs, and Black Falcons glide on flat or drooped wings. Though plumage colour varies considerably with sex and age, from uniformly dark like a Black Falcon at a distance (close-up some Black Falcons have narrow basal barring under the primaries, and faint bars under the tail) to pale like a Nankeen Kestrel, dark Brown Falcons have barred underwings and tail, which Black Falcons lack. Brown Falcons can also be confused with Swamp Harriers, but Brown Falcons have more tapered, less upswept wingtips, and they lack the harrier's pale rump.

Brown Falcons often sit upright on dead trees, fence posts, power poles or wires. They can form loose flocks around plentiful food sources, especially insects and rodents. They are excitable, scrappy, highly vocal, and present in most open habitats but avoid dense forest. They fly with a deep rowing action, with heavy, slow wing beats, often erratic with jinking and side-slipping. They are fast in pursuit of avian prey and when attacking enemies, usually other raptors and often other Brown Falcons, they are capable of breath-taking stoops. They lack the stiff flickering wing beats of other falcons.

Habitat: Diverse. They occur in all habitats, but are more common in open country and scattered woodland.

Territoriality and behaviour: Brown Falcons are the most commonly seen raptor in roadside counts across Australia, often seen perched on fence posts, bushes, utility poles or sometimes balancing on powerlines with their tail fanned. It is the noisiest falcon and commonly cackles while flying, not unlike a laying hen. At the start of the breeding season males soar and perform erratic diving displays rocking from side to side, flashing their breast and underbelly, with flickering flight and cackling calls. The pair soars together with laboured wing beats, squawking and cackling. Males greet females with a bowing display and chattering calls, females respond and they may allopreen. Males present food that females eat or cache. Pairs are extremely territorial against all raptors but especially against other Brown Falcons.

Some Tasmanian Brown Falcons are said to migrate to the mainland, and some southern mainland birds are said to migrate to the tropics.

Food and hunting: Very diverse diet, possibly the most diverse of any Australian raptor; opportunistic carnivores that take a wide range of prey types including mammals, birds, reptiles, insects, aquatic animals such as molluscs and snails, rarely fish, and carrion (Figure A1.19). Around the ACT the diet is also diverse, and shows strong variation between territories; the main items are insects, small mammals, rabbits, birds, skinks and molluscs. They often drop from telephone poles, fence posts, wires or treetops onto prey in a slanting glide or stoop, parachute down onto terrestrial prey, prospect by soaring, or direct flying attack that might become a short chase. They sometimes pursue prey at high speed, often in low fast and/or twisting flight. Usually they seize prey on the ground.

They also run after vertebrate prey on the ground, atypical of most falcons, pursue insects on foot and rob other raptors. They also follow fires, livestock, other animals and farm machinery and take flushed prey. They are expert snake killers.

Breeding: Laying season is April to September in northern Australia and August to October in central Australia and the south. Some Brown Falcons in the high country lay eggs in August but most lay in September and some in October (Figure A1.20), usually in an old corvid nest but also in the nest of another raptor, including Wedge-tailed Eagles, and usually situated in a tree, 4–30 m above the ground but also on a power pylon.

Age at first breeding said to be 3 years for males, 2 years for females. The oldest banded bird was 18 years old; longevity in captivity is up to 20 years.

Eggs: The clutch is most often three, but ranging from one to five, pale buff or buffy-white, splotched with reddish to purplish-brown, incubated for ~32–35 days.

Young: Young hatch in late September or October and fledge ~36–42 days after hatching. The nestling has a distinctive pale rufous first down, not white as in most

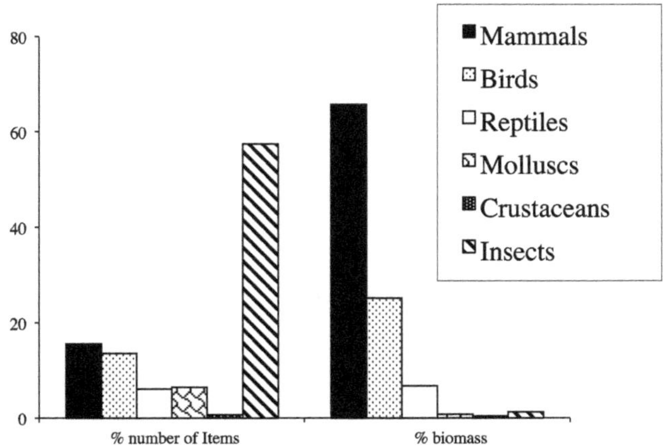

Figure A1.19 Diet of Brown Falcon in the ACT region (Fuentes, Olsen and Rose unpublished).

Mar April May June July Aug Sept Oct Nov Dec Jan Feb

I——I
Pre-breeding
I——I
Eggs
I——I
Hatching
I————I
Nestlings
I————I
Fledglings

Figure A1.20 Approximate timing of breeding of Brown Falcons in the high country.

falcons, and pale grey second down. Various measures of breeding success include 69% fledging (of eggs laid), 1.77 young fledged per nest and 2.39 per successful nest, 2.3–2.4 young per successful nest and 1.7 per territorial pair. Newly independent young often feed on invertebrates such as crickets and grasshoppers and, in autumn, they can be seen in loose flocks of up to 50 birds. Fledgling dependence on parents is said to last up to 6 weeks, then young disperse widely, some up to 2000 km.

Voice and calls: Brown Falcon calls differ from those of other falcons. They give a chicken-like rattling cackle, the most commonly heard call, also clucks and a single note *kake*, and food-begging whine.

Status: Not considered threatened globally or nationally. It is adaptable and has benefited from habitat modification for farming and agricultural activities and introduced species such as starlings and rabbits.

Myths and questions: Long considered to have several colour morphs unrelated to sex and age, a myth perpetuated because of an incorrect claim in Marchant and Higgins (1993) that captive birds remained unchanged for several years. Weatherly *et al.* (1985) and McDonald (2003a) linked different colour phases to ages and sexes of individuals but these questions still remain open.

Sources: Debus (2012); Olsen and Fuentes unpublished data; Bollen (1993); Marchant and Higgins (1993); McDonald (2003a,b, 2004); McDonald *et al.* (2003); Olsen (1992); Aumann (2001); McDonald *et al.* (2012).

(Julian Robinson)

Australian Hobby

Falco longipennis

The specific name *longipennis* is combined Latin for *longus* (long) + *penna* (feather, fin or wing) for this falcon's long wings and long primary feathers. Hobbies are a group of small Eurasian/Australasian falcons comprising four species: the Australian, Oriental, African and Northern Hobby.

Field recognition

Signature characters: Small, slender, dark falcon with dashing flight and elegant sickle-shaped wings; in flight, their shape similar to a White-throated Needletail. Also like a miniature Peregrine with slightly longer wings. **Adults** have an incomplete black helmet with light patch behind the eye, grey back and tail, and rufous breast. **Juveniles** browner. Males are the size of a Crimson Rosella or Kestrel and females the size of a Feral Pigeon. At rest they show a slim build, small beak, medium length legs, and long wings reaching to or just beyond the tail-tip.

Iris dark brown, legs and feet dull yellow.

It is 30–37 cm long (the tail makes up almost half this length), with a wingspan of 66–90 cm. The average male weighs ~220 g, females ~290 g.

When hunting or displaying, its flight action is dashing, flickering, and quick, cutting over or between trees in acrobatic pursuit of small birds, bats and flying insects, but sometimes it can be more leisurely. It glides on flat or slightly drooped wings and soars on pointed, flat wings. The Australian Hobby is most often confused with the kestrel, or with the Peregrine Falcon, which is more heavily built and a more powerful flier with a heavier head and beak, full dark helmet, shorter, broader more triangular-shaped wings, shorter tail and heavier feet. Pale juvenile inland Hobbies sometimes stray into Canberra and these can be confused with Nankeen Kestrels which lack the dark head markings and rufous underparts. The Nankeen Kestrel is paler, with a longer round-tipped tail and less sharply pointed wings. The Collared Sparrowhawk is also about the same size but has shorter, broader, rounded wings that don't reach the tail-tip, and longer legs.

Habitat: Mainly found in lightly timbered country, dry open forests, water courses, and open areas with scattered trees and small woodland patches. They are adaptable and can breed on grazing land, in mountains and cities, but they tend to avoid cliffs and escarpments where Peregrines breed, apparently because Peregrines displace them. However, they nest in heavily forested regions such as Namadgi National Park where they hunt birds and insects over the forest canopy and over dams.

Habits: Australian Hobbies are typically seen dashing over or past, or perched on a high dead branch, television aerial, wire or pole. They hunt well into dusk, even after dark, even over the city.

As with most raptors, the smaller male does most of the hunting while the larger female guards the nest from currawongs, ravens and other enemies. After young are half-grown, the female often begins hunting, sometimes close to the nest or fledged young. Males pass food to females in midair or at high perches. Males and females perform a bowing display on the nest, with loud *chip-chip* and chittering calls. Some birds breeding at high latitudes and high altitudes, such as the high country, leave in autumn, returning in spring; there is a corresponding winter influx to the tropics and we have seen them in Indonesia during the austral winter.

Food and hunting: They cruise over the tree canopy like a small terrain-hugging missile, surprising small birds or bats and flying insects; they sometimes fly sortie after sortie from a perch at passing flocks of parrots or passerines. They almost never land on the ground, preferring instead to snatch prey in midair and eat it on the wing or on a high perch. They hunt by day, well after sundown, and sometimes at night by artificial light. They also forage by high, fast flight, or still-hunting from a prominent perch. The birds taken are mostly around 75 g, though they can catch animals up to 250 g. Mostly they take terrestrial flock species including grass finches, doves and small parrots but they also take larks, pipits and aerial species such as swallows (Figure A1.21). Near Canberra they catch many Common Starlings and House Sparrows, and aviary escapees such as Budgerigars. Females sometimes catch prey as large as rosellas and, like all falcons, kill the prey by snipping the spinal cord with a bite, to avoid being bitten themselves.

Breeding: In south-eastern Australia they breed in spring and summer (Figure A1.22); in the high country they breed later than most other raptors. As with all falcons, they cannot build a nest so they use the nest of a raven after the earlier nest owner has finished that spring – that is, they seem to prefer a 'fresh' nest, not a nest from previous years. To date, it seems nobody has tried artificial nests for hobbies, but the closely related Eurasian Hobby does use basket/stick nests set in trees so Australian Hobbies may do the same. Typically the corvid nest used is high, more than 10 m above the ground, in the top of a tree or on an electricity pylon, but they will also use the nest of a hawk, eagle or kite.

Eggs: Two to four eggs, usually three, pale-buff or ground coloured spotted or blotched with rufous or darker brown are laid in early November and incubated for ~30 days (28–35 days).

Nestling: White to cream coloured down, or sometimes pale rusty-brown. In the high country nestlings usually hatch in December, stay in the nest for 34–41

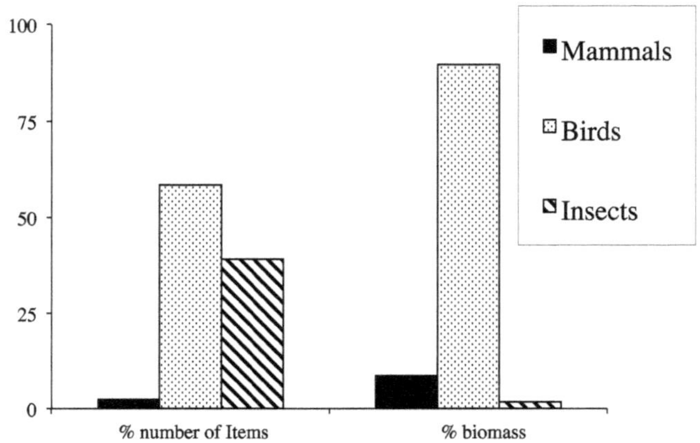

Figure A1.21 Prey of Hobbies nesting in the ACT region (Fuentes, Olsen and Rose unpublished).

Mar April May June July Aug Sept Oct Nov Dec Jan Feb

I——I
Pre-breeding
I——I
Eggs
I——I
Hatching
I————I
Nestlings
I——I
Fledglings

Figure A1.22 Approximate timing of breeding for Australian Hobbies in the ACT region.

days and fledge in January or early February. Young stay in the nest area for about 1 or 2 months after fledging. Success near Canberra was 2.75 young per year for a pair over 4 years; successful broods comprising 2–3 young. The period of dependence after fledging lasts up to 3 months, after which young disperse or migrate widely; up to 900 km has been recorded. The oldest recovered banded bird was 7 years old.

Voice and calls: Its most commonly heard call is a sharp *cac-cac-cac* used when attacking enemies, higher pitched than a Peregrine Falcon, and similar to a Nankeen Kestrel or Brown Goshawk but reedier, with more timbre. They also give chittering calls and *chip-chip-chip* calls to mates, and males give a loud chuckling call in a winnowing pre-breeding display flight.

Status: The Australian Hobby is not globally or nationally threatened. It is fairly common and widespread, including in cities and towns and its population seems stable. It may have benefited from introduced sparrows and starlings. It seldom bothers aviaries or takes pigeons and is seldom shot.

Myths and questions: Subspecies may not be valid; reports that hobbies take ducks or Galahs are almost certainly misidentified, taken instead by Peregrines.

Sources: Debus (2012); Debus *et al.* (1991); Metcalf (1989); Olsen and Fuentes unpublished data; Marchant and Higgins (1993); Olsen (1992); Olsen *et al.* (2008a); Aumann (2001).

Peregrine Falcon

(Julian Robinson)

Falco peregrinus

The Latin *Falco* refers to sickle-shaped bill and talons; *peregrinus* is Latin for 'strange' or 'foreign' but has come to mean 'wandering' for Peregrines because some of the nominate *F. p. peregrinus* in Europe migrate. Adults of the Australian race *F. p. macropus* (large feet) are sedentary, though juveniles may wander.

Field recognition

Signature characters: The **adults**: dark blue-grey back, light fronted raven-sized (female) or magpie-sized (male) falcons with triangular-shaped, pointed wings, direct flight on quick wing beats with no hovering, occasional soaring, characteristic black helmet and white bib, with black barring on the light-coloured breast.

In flight wings are triangular or anchor-shaped, depending on wind and wing angle. **Juveniles** are browner with dark teardrops on the breast.

Female Peregrines weigh ~790 to 1100 g (average 920 g, $n = 12$); males weigh ~538 to 675 g (average 610 g, $n = 4$). Wingspan is ~850 mm. Desert birds can be paler, some coastal Peregrines and Tasmanian birds especially are darker.

Habitat: Peregrines around Canberra inhabit most types of habitat that have cliffs, but no Peregrines in Australia have been found nesting in or above the subalpine zone. Mooney and Brothers (1988) consider 900 m the altitudinal limit in Tasmania; in the Brindabellas and Kosciuszko National Park it is 1200 m. This must relate to a lack of prey or severe weather, because Peregrines at similar latitudes in North America and elsewhere nest twice as high, and the Australian limit is similar to that of southern Scotland. Unlike Peregrines at higher latitudes, they appear to be non-migratory.

Territoriality and behaviour: Very aggressive at the nest, attacking and striking intruders, especially humans or other raptors. Pre-breeding they often soar and perform acrobatic dives. Sherrod (1983) watched two nests in Victoria and tallied the number of attacks made by adults, against – humans (18), magpies (9), Brown Falcons (7), Australian Ravens (4), Feral Dogs (3), Sulphur-crested Cockatoos (3) and Black Cormorant, Brown Goshawk, Red Fox, unidentified kite or eagle (1) each. Scavengers regularly combed the cliffs for prey and Sherrod saw magpies take cached prey even with adults present.

Food: Peregrines prospect using contour-hugging flights and soaring. They are excellent bird hunters and catch birds of almost every size, from wrens to ducks, in midair. They catch birds in a near vertical stoop clocked at up to 190 km/h, in a long slanted stoop, or in a tail chase. Near Canberra they appear to have adapted to hunting Galahs and smaller parrots, but they also eat introduced Domestic Pigeons and Common Starlings (Figure A1.23). Given their body size, Australian Peregrines have relatively larger beaks than many of their overseas counterparts because the neck vertebrae of their prey, such as Galahs, are thicker than are the vertebrae of similar-sized pigeons and other prey birds, and Galahs bite back so they have to be killed quickly.

Breeding: No falcons build their own nests. Most known ACT nests are on cliffs, buildings, or in abandoned and active quarries, though there is a population using old eagle and corvid stick nests at Lake George in New South Wales.

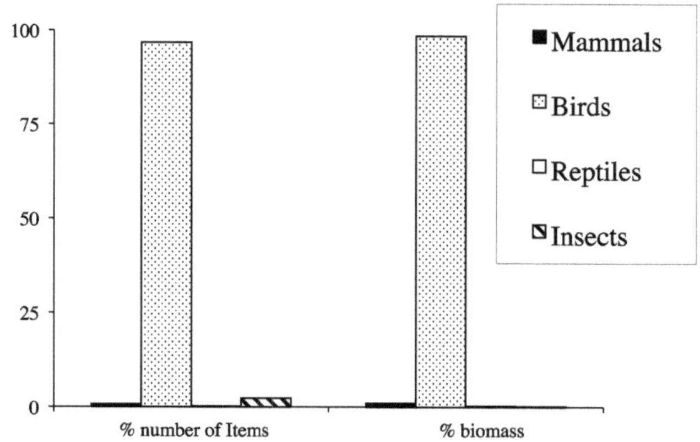

Figure A1.23 Peregrine diet in the high country (Fuentes, Olsen and Rose unpublished).

Peregrines occupy suitable cliffs for many generations and, at one Tasmanian Peregrine site, carbon dating extends occupancy back 19 000 years. Peregrines on rivers near Canberra nest 2 km apart, starting around August (Figure A1.24), where there are abundant suitable cliffs. The density is remarkably stable, with each territory having a resident pair every year and no new pairs coming in or leaving, except where quarries create new nest sites or where Peregrines take up ledges on buildings. Age at first breeding is said to be 2 years for females and 3 years for males.

Eggs: Two to five eggs, usually three, laid in late August–early September, pale buff heavily blotched with reddish-brown and blackish. Eggs hatch after 35 days incubation.

Young: Nestlings are white, with a second greyer down as they age. Later, dark primary and tail feathers poke through the light down making a contrasting white and dark-brown nestling. Around Canberra, Peregrines fledge 1–4 young in November and early December after ~42 days in the nest. In south-eastern

Mar April May June July Aug Sept Oct Nov Dec Jan Feb

I——I
Pre-breeding
I——I
Eggs
I——I
Hatching
I————I
Nestlings
I——I
Fledglings

Figure A1.24 Approximate timing of breeding for Peregrines in the high country.

Australia 58% of nests fledge young, depending on rainfall (lower productivity in high rainfall years), ~1.23 young fledged per territory, 2.16 per successful nest (brood size). In Tasmania ~64% of pairs are successful; in Victoria productivity is 1.3–1.57 young per pair, 2.0 per successful pair and 1.3–1.55 per clutch started. Period of dependence after fledging is said to be ~2 or 3 months, but this needs further study, as young can linger in the natal territory for up to 8 months; one individual dispersed 500 km from its nest, but males usually disperse ~60 km, females ~130 km. Annual survival is 36% for juveniles, 94% for adults, with longevity in the wild up to 14 years.

Voice: Loud repeated *cac-cac-cac* when attacking. Other calls including food-whining *waik-waik-waik* from both males and females and *eechip-eechip* calls that females especially use when communicating with their mates.

Status: Not considered threatened globally or nationally. Peregrines were heavily affected by DDT use in America and Europe, but populations have now recovered there. The main threats are disturbance at nest sites and illegal taking of eggs or young, which can affect reproductive success. The local breeding population has been very stable over the last 25 years.

Myths and questions: Claims of skewed sex-ratio of nestlings were refuted by Krackow (1993).

Sources: Debus (2012); Olsen and Fuentes unpublished data; Marchant and Higgins (1993); Olsen (1992); Olsen (1995); Olsen and Rehwinkel (1995); Olsen and Tucker (2003); Olsen and Olsen (1988); Olsen *et al.* (2004, 2006a, 2008a); Emison *et al.* (1998).

Appendix 2 Sick and injured raptors

See *Caring for Birds of Prey* (Olsen 1990) for a more thorough treatment of this topic.

Picking up a sick or injured raptor

Most sick or injured raptors found in the high country, including ACT suburbs or along roadways, are collision-victims. A sick or head-injured raptor will sit drooping its head and looking fluffed up. In contrast, an otherwise injured raptor is sleek, until you get close when it may puff its feathers as a threat. Fast-flying bird-catchers, such as falcons or accipiters, are over-represented in rehabilitation facilities, apparently because they more often collide with obstacles. Others, such as Southern Boobooks, are often hit by cars, and both Southern Boobooks and Barn Owls receive head injuries from attacking magpies and currawongs.

Minimising stress to the raptor is your first priority. Wild birds are often killed by shock rather than the injuries they receive. If an adult raptor allows you to approach it, it may be in shock or badly injured. Don't handle the bird too much and further stress it. Concentrate on how to secure the raptor.

Raptors are powerful animals. Even Southern Boobooks and kestrels have astounding power in their feet and talons; larger raptors have even more. When held by humans, raptors can struggle, break feathers, hurt themselves, or use their powerful beaks and claws to injure the humans trying to help them. Raptors flip onto their backs and present their talons, their most dangerous weapons, to hurt their enemies; it is important that you pick up the raptor carefully. One procedure is:

1 if the raptor is on its back defending itself, toss a towel over it, or use some fabric tossed towards it to engage its claws;

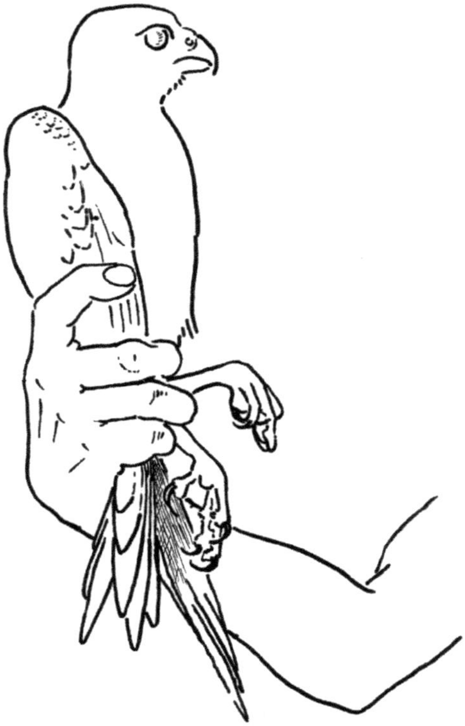

Figure A2.1 Holding a raptor so its talons don't entangle or hook into the carer.

2 place a hand in the middle of the raptor's back and press down and forward so that its legs extend backwards out under its tail;
3 encircle the body, including both wings and legs, by wrapping your hands around the raptor's lower back just above where the tail feathers emerge;
4 pick it up with its beak facing away from you and carefully disengage claws from the netting or foliage. Place a finger between the raptor's tarsi so it does not grab its own feet (Figure A2.1).

A biologist working long enough with owls, hawks, falcons or eagles will be grabbed by the raptor he or she is trying to hold. The raptor slashes out with needle sharp talons at any part of the biologist that strays into range, and the grip is powerful and locks on because of a tendon mechanism, so talons are hard to detach.

Remain vigilant. Raptors such as owls seem to have short legs, but this is an illusion. They have long legs, long tibiae and tarsi. The legs look short because their body and flank feathers (trousers) hang down over the top half of their tarsi. Raptors such as Southern Boobooks or accipiters reveal their long legs when they hunt. They extend their legs with their feet wide open and hit the tree canopy, or chase birds into the tops of bushes and grab them with their outstretched legs and toes. While being held, a raptor's long legs can shoot out, grab you, and puncture you with four talons.

If this happens, follow a few simple rules:

1 If alone, hold the raptor firmly so it cannot free the other leg and grab you with the other foot.
2 Do not make jerky movements that can hurt the raptor, or dig the talons in deeper into you. Stay calm.
3 You, or an assistant, need to straighten the raptor's leg, not by pulling the leg out, but, instead, by grasping the femur, tibia and tarsus, and gently forcing the bends out of the raptor's leg so the locking tendons that hold the four grasping talons curled into you are released. This tendon arrangement gives the raptor its powerful grip and allows the raptor to sleep on a perch without falling off. By understanding this locking mechanism, you can use it to your benefit.
4 With the raptor's leg straightened, remember that talons are deeply curved, not straight, and each talon is concave underneath, not tubular (except for Black-shouldered Kites). That is, with the leg straightened, if you pull the talons straight out of you, they will pull out your flesh with them. Carefully pull out each talon around the curve of its shape, one at a time.
5 Disengage each talon from your clothing.

Cages and bandaging

Never put a raptor in a wire cage. Wire damages the feathers and cere and makes the raptor unreleasable. Place the raptor in a darkened cardboard box perforated with holes along its sides near the bottom, and lined with newspaper or a towel. Do not use straw or sawdust, and do not place water in the box. Do not try to feed the raptor; keep the raptor cool.

If a raptor's wing is damaged and hanging, you can sometimes use a Figure 8 bandage to hold it in place (Figure A2.2).

A broken tarsus can be splinted (Figures A2.3, A2.4).

Take the raptor to your local wildlife carer/rehabilitation group (e.g. WIRES in NSW) or to the RSPCA. If you do a lot of raptor rehabilitation, keep hoods available as they sedate the bird and prevent it from injuring itself (Figure A2.5).

If you keep raptors for any length of time, particularly for captive breeding, consider properly designed non-wire pens (Figure A2.6).

Returning raptors to the wild is rewarding, a way of redressing the balance. After all, most were injured by humans or human-built structures in the first place. Rehabilitation puts us close to a wild creature for a time (Figure A2.7).

Figure A2.2 Figure 8 bandage to hold an injured wing in place.

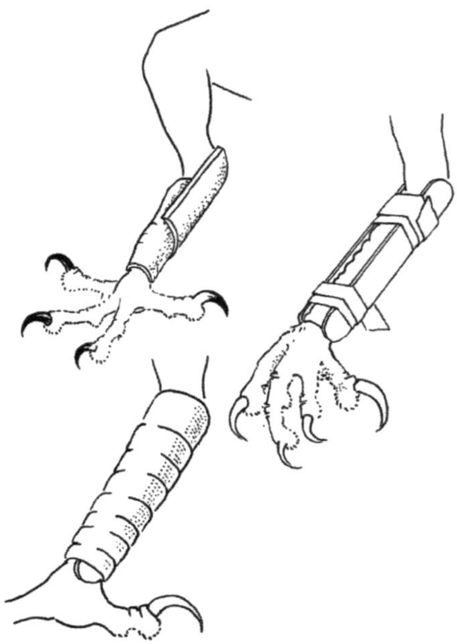

Figure A2.3 Splinting a broken tarsus.

Figure A2.4 Brown Goshawk with splinted and bandaged broken tarsus.

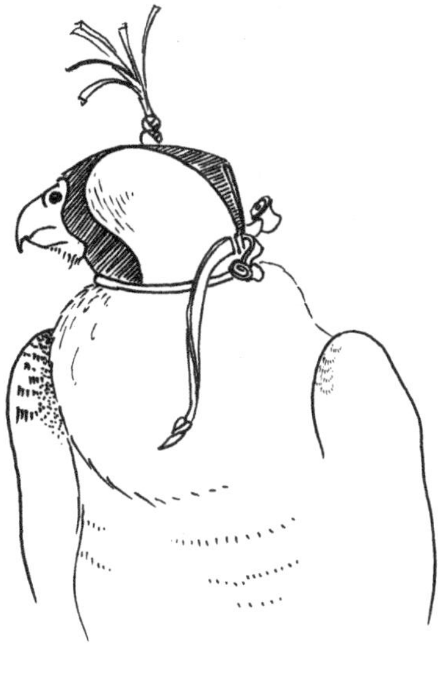

Figure A2.5 Hooded falcon.

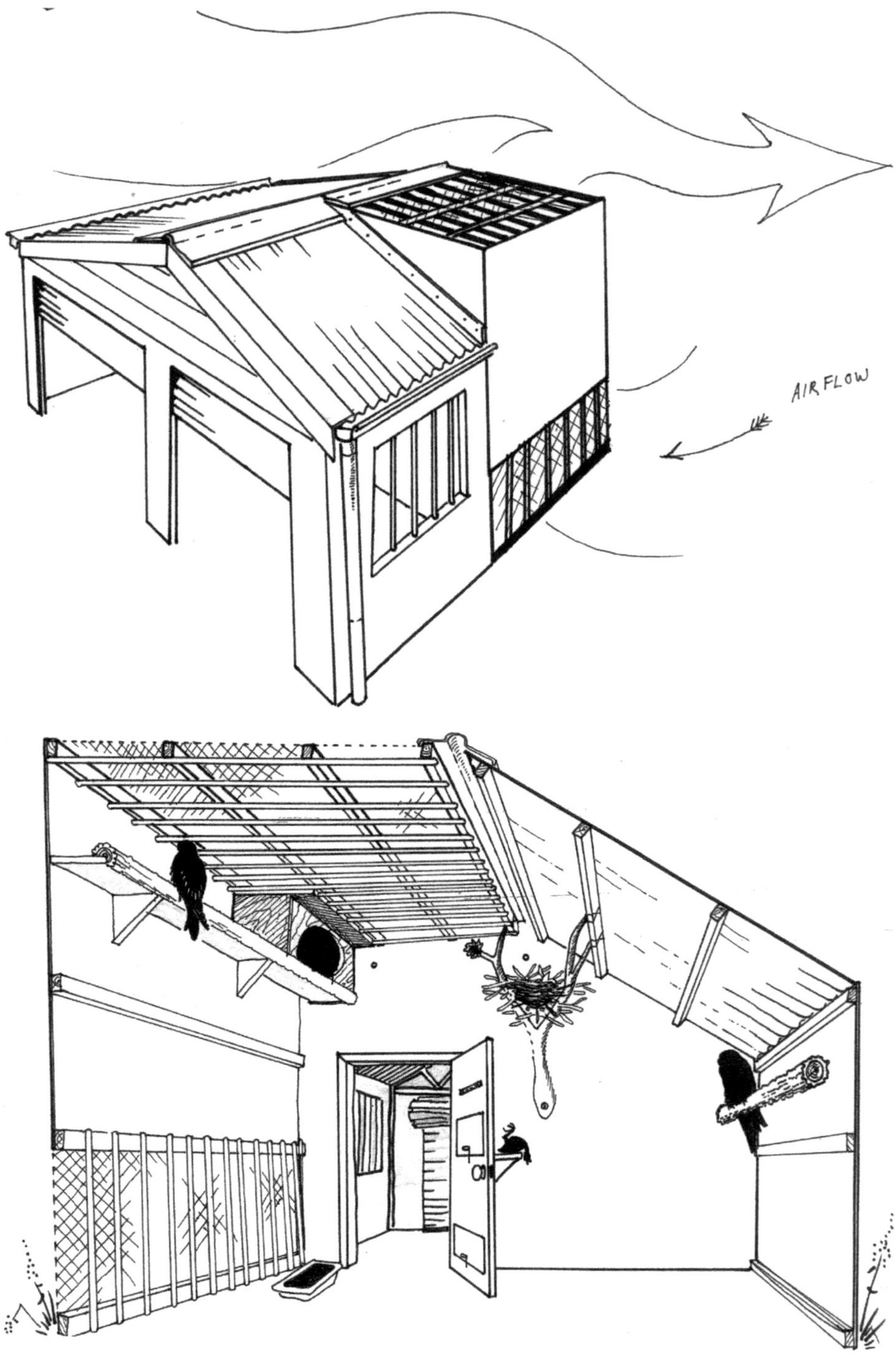

Figure A2.6 Holding pen design for keeping raptors.

Figure A2.7 Colin MacLaren releasing a previously injured Peregrine.

Scientific names of raptors mentioned in the text

Turkey Vulture *Cathartes aura*
California Condor *Gymnogyps californianus*
Eastern Osprey *Pandion cristatus*
American White-tailed Kite *Elanus leucurus*
Eurasian Black-winged Kite *Elanus caeruleus*
Black-shouldered Kite *Elanus axillaris*
Letter-winged Kite *Elanus scriptus*
Black Kite *Milvus migrans*
Whistling Kite *Haliastur sphenurus*
Brahminy Kite *Haliastur indus*
White-bellied Sea-Eagle *Haliaeetus leucogaster*
Bald Eagle *Haliaeetus leucocephalus*
White-tailed Sea-Eagle *Haliaeetus albicilla*
Spotted Harrier *Circus assimilis*
Swamp Harrier *Circus approximans*
Brown Goshawk *Accipiter fasciatus*
Grey (White) Goshawk *Accipiter novaehollandiae*
Collared Sparrowhawk *Accipiter cirrocephalus*
Cooper's Hawk *Accipiter cooperii*
Eurasian Sparrowhawk (Northern Sparrowhawk) *Accipiter nisus*
Eurasian Goshawk (Northern Goshawk) *Accipiter gentilis*
Gray Hawk *Buteo nitidus*
Red-shouldered Hawk *Buteo lineatus*
Swainson's Hawk *Buteo swainsoni*
Red-tailed Hawk *Buteo jamaicencis*
Rough-legged Hawk (Rough-legged Buzzard) *Buteo lagopus*
Spanish Imperial Eagle *Aquila adalberti*

Golden Eagle *Aquila chrysaetos*
Wedge-tailed Eagle *Aquila audax*
Bonelli's Eagle *Aquila fasciata*
Little Eagle *Hieraaetus morphnoides*
Brown Falcon *Falco berigora*
American Kestrel *Falco sparverius*
Eurasian Kestrel (Common Kestrel) *Falco tinnunculus*
Nankeen Kestrel *Falco cenchroides*
Merlin *Falco columbarius*
Eurasian Hobby (Northern Hobby) *Falco subbuteo*
Australian Hobby *Falco longipennis*
New Zealand Falcon *Falco novaeseelandiae*
Prairie Falcon *Falco mexicanus*
Laggar Falcon *Falco jugger*
Lanner Falcon *Falco biarmicus*
Saker Falcon *Falco cherrug*
Gyrfalcon *Falco rusticolus*
British Peregrine Falcon *Falco peregrinus peregrinus*
Australian Peregrine Falcon *Falco peregrinus macropus*
Sooty Owl *Tyto tenebricosa*
Masked Owl *Tyto novaehollandiae*
Common Barn Owl *Tyto alba*
Eastern Barn Owl *Tyto javanica*
Great Horned Owl *Bubo virginianus*
Boreal Owl *Aegolius funereus*
Rufous Owl *Ninox rufa*
Powerful Owl *Ninox strenua*
Barking Owl *Ninox connivens*
Southern Boobook *Ninox novaeseelandiae*
Christmas Island Hawk-Owl *Ninox natalis*
Little Owl *Athene noctua*

Glossary

Accipiter Genus of near cosmopolitan, mostly woodland and forest, hawks (goshawks and sparrowhawks) characterised by short wings and long tail.

Accipitridae/Accipitriformes Family and order of hawks, kites, eagles and Old World vultures.

Aerial Capable of flight, as opposed to **arboreal** or **terrestrial**.

Agonistic Attack and fleeing behaviour, related to nest and territory protection, and social dominance and submission.

Allopatry When two or more populations occupy completely separated geographic areas, at least during breeding, for example, Whistling Kites and Powerful Owls in the ACT.

Allopreen When some raptors such as owls and eagles preen (groom) a conspecific, usually their mate or sibling, often the partner's head, in pair bonding.

Altricial Young hatched blind and relatively helpless, that is, all raptor young (versus **nidifugous**).

Alula Cluster of miniature flight feathers at the wrist area used for braking and slow flying.

Ambush hunting Remaining concealed on a cliff, in a tree, or other perch waiting for prey to appear.

Anisodactyl Avian foot arranged with three toes forward and one back (all diurnal raptors except Osprey) (see also **Zygodactyl**).

Arboreal Frequenting trees as opposed to **aerial** or **terrestrial**.

Arthropod An invertebrate with hard exoskeleton and jointed legs such as an insect, spider or crab.

Bandicoot Small ground-dwelling marsupial of the family Peramelidae, never recorded in the ACT, but caught by eagles there.

Biomass Total weight of organisms taken as prey or in a given area of land.

Bloom A coating of fine dust, often greyish, especially on the back of some raptors, produced by specialised powder on down feathers.

Booted Fully feathered tarsus, as in 'booted eagles'.

Brancher A nearly grown diurnal or nocturnal raptor that has left the nest but cannot fly properly.

Brooding Parent sitting or crouched over young, resting somewhat higher than when they are incubating.

Brood patch Patch of denuded skin on the belly of many birds – mainly female, rich in blood vessels that supply heat during incubation and brooding. Helpful for sexing certain owls.

Cache To hide surplus killed prey in crevices, nests, tree forks, etc. for later retrieval.

Cain and Abel struggle A larger, usually older sibling kills or evicts a younger sibling from the nest.

Casting Oblong packet of indigestible material such as fur, feathers, bones or insect casings regurgitated by raptors and some other birds such as currawongs (synonymous with *pellet*).

Cere Soft skin at the base of the bill, in which the nostrils of a raptor are located.

Cone cell A specialised light receptor in the retina that perceives colour.

Congeneric Members of the same genus.

Conspecific Members of the same species.

Contour-hunting Flying at speed just above the ground or canopy, following dips and folds of the land, aimed at surprising prey.

Corvid Bird in the family Corvidae; in Australia this means crows and ravens in the genus *Corvus*.

Crepuscular Active at dawn and dusk.

Currawong Crow-like, black, or grey-and-white bird of the genus *Strepera*, related to Australian Magpies and butcherbirds *Cracticus* in the family Artamidae.

Dasyurid A carnivorous marsupial such as quoll or antechinus in the family Dasyuridae.

Dichromatic Two plumage forms within a species, for example, Nankeen Kestrel.

Dimorphic Species with two physical forms, for example, sexual dimorphism where male and female are of different size or plumage.

Diurnal Active by day.

Dorsal Pertaining to the back or upperparts of a bird, as opposed to **ventral**.

Ecotone Boundary or mixing zone between two vegetation or habitat types, for example, between forest and grassland.

Ectoparasite A parasite living on the body surface as opposed to an **endoparasite** living inside the body.

Endemic Unique to a particular region.

Endoparasite A parasite living inside the body as opposed to an **ectoparasite** living on the body surface.

Exoskeleton Hard armour-like covering on arthropods that waterproofs and protects them.

Falconidae/Falconiformes Family and order that includes falcons and caracaras.

Fledge First flight from the nest.

Flying-fox A fruit-bat, family Pteropodidae.

Floater An independent bird lacking a breeding territory or mate.

Genotype The genetic basis for an organism's traits.

Genus Taxonomic category containing species believed to be closely related.

Glider A gliding possum of the genera *Petaurus* (family Petauridae), *Petauroides* (family Pseudocheiridae) or *Acrobates* (family Acrobatidae).

Hacking Process of allowing juvenile captive raptors unlimited food while they develop the skills of foraging.

Harrier Vernacular for hawks in the genus *Circus* noted for unusually long wings and tails, long legs, and owlish-shaped faces and facial discs.

Hawking Catching insects or other prey in midair.

Hectare 10 000 square metres (2.47 acres).

Hobby Vernacular for a group of small, long-winged, part-insectivorous falcons (*Falco*).

Home range Area regularly used but not necessarily defended by a raptor over a defined period (see also **Territory**).

Hover Form of flapping flight where the raptor remains stationary while beating the wings rapidly.

Incubate Warming a clutch of eggs by lying or sitting on them (see **Brooding**).

Interspecific Occurring between different species.

Intraspecific Occurring within a single species.

Iris The coloured part of the eye, surrounding the pupil; plural is irides.

Irruption Sudden dramatic movement of a species from one area to another; also, a dramatic explosion of numbers by prolific breeding *in situ*, for example, the House Mouse.

Juvenile A bird in its first (juvenal) non-downy plumage usually obtained, at least in part, before leaving the nest.

Kiting Form of flight where the raptor remains stationary by directing itself into a strong wind with little or no heavy beating of the wings.

Kleptoparasitism Piracy; robbing another bird or animal of food.

Kneading A hawk or eagle convulsively grips the body of prey trying to kill it, often using alternate feet.

K-selected Reproductive strategy of species said to usually be large, with low reproductive rate, in predictable environments, long-lived, with stable populations close to the carrying capacity (K) of density dependent controls (versus **r-selected**).

Malar stripe or patch A dark stripe in falcons extending down from the lip or eye.

Mantle The raptor hides prey with its spread tail and wings while often raising its back and crest feathers.

Mobbing Harassing a predator to drive it away.

Monotypic A taxonomic category with only one member, as opposed to **polytypic**; a monotypic genus contains a single species.

Morph A lifelong plumage type, unrelated to age or gender (as opposed to 'phase' which is a transitory, age-related condition).

Nares Nasal cavities opening via the nostrils.

Nestling Unfledged bird still in the nest.

Nestling period Length of time from hatching to fledging (first flight).

Niche The sum of structural, physiological and behavioural adaptations of a species to its environment; the full range of environmental conditions within which it can survive.

Nictitating membrane Third eyelid in reptiles and birds that sweeps over the eye from the inside corner to clean and protect it.

Nidicolous Species whose **altricial** young stay in the nest until fledging, or nearly so (all raptors), versus **nidifugous**.

Nidifugous Species whose precocial young leave the nest soon after hatching such as quail (not raptors), versus **nidicolous**.

Nocturnal Active by night, as opposed to **crepuscular** or **diurnal**.

Nomadic A species or individual whose movements are relatively unpredictable in timing, direction and duration.

Orthopteran Insect belonging to the order of grasshoppers and crickets (literally means *straight-winged*).

Passerine Bird in the order Passeriformes characterised by four toes arranged at the same level, the hallux directed backwards and non-reversible, well adapted to perching, and toes are never webbed. They can be small, like fairy-wrens, or larger like ravens. Other groups, including waterbirds, sea birds, raptors and others, are commonly grouped as *non-passerines*.

Pellet See **Casting**.

Phenotype The appearance of an organism, irrespective of the genetic basis of its appearance.

Photoperiod Length of day.

Polytypic A taxon containing more than one member, for example, *Accipiter*.

Possum In Australia, an arboreal marsupial of the Superfamilies Petauroidea and Phalangeroidea, typified in the high country by the Common Brushtail Possum *Trichosurus vulpecula* and Common Ringtail Possum *Pseudocheirus peregrinus*.

Preen Clean, oil and arrange the feathers with the beak.

Primaries Outer flight feathers attached to the bones of the hand and digits; with **secondaries** they make up the **remiges**.

Quartering Flying low to the ground over large areas, usually at slow speed searching for prey.

Quoll A carnivorous marsupial in the genus *Dasyurus*.

Raptor A bird of prey, typically with powerful feet, sharp talons, and a sharp, curved beak.

Rectrices Tail feathers, singular *rectrix*.

Remiges Primaries and secondaries, also called flight feathers, singular *remex*.

Reversed sexual dimorphism In animals where the female is larger than the male; found in most raptors, but the reverse of what is found in most birds.

Rictal bristles Small bristle-like feathers of variable length near the gape and cere, often covering the nostrils, said to have some sensitivity in raptors such as owls.

Riparian Pertaining to the banks of rivers, dams and streams.

Rod cell A light receptor in the retina involved in black and white vision.

Rodenticide A poison for killing rodents (rats and mice).

r-selected Reproductive strategy of species said to have a high reproductive rate, in unpredictable environments, shorter-lived, with markedly fluctuating populations, good powers of dispersal; r is the technical notation for Intrinsic Rate of Increase of a population (versus K-selected). In the high country, Barn Owls and Black-shouldered Kites would be toward the r-selected pole of the continuum, Wedge-tailed Eagles and Peregrines towards the K-selected pole of the continuum.

Secondaries Inner flight feathers (remiges) attached to the ulna.

Secondary poisoning Non-target poisoning of an animal after it consumes a target animal such as a rabbit.

Soar-hunting Finding prey while soaring.

Strigidae 'Typical owls', including all but the **Tytonidae.**

Still-hunting Hunting from a perch or while sitting on the ground waiting for prey.

Subspecies Taxonomic category of recognisable geographic division of a species, for example, *Falco peregrinus macropus* (genus, species, subspecies); also called *race.*

Sympatric Two or more species or populations occupy the same area, especially during breeding.

Tarsus (plural tarsi) The lower half of the visible part of a bird's leg (actually the foot) segments directly above the toes. The true knee is hidden by the flank feathers. In owls and 'booted' eagles it is feathered, in other raptors bare.

Taxon A taxonomic unit of any rank, for example family, genus, species, subspecies.

Terrestrial Associated with the ground as opposed to **aerial** or **arboreal**.

Territory Defended whole or part of a **home range**, centred on the nest when breeding.

Trichomoniasis A potentially fatal and infectious disease common in Columbiformes (pigeons) caused by the flagellate protozoa *Trichomonas gallinae* and transmitted to raptors that eat infected pigeons.

Trophic level A step in the movement of energy through the ecosystem, for example, grass to rabbits to eagles are three trophic levels.

Tytonidae Barn Owls and Grass Owls, one of the two families of owls (for the other family see **Strigidae**).

Ventral Underside, opposite of **dorsal**.

Vermiculated Marked with fine, wavy ('worm-like') lines.

Wallaby A medium-sized kangaroo (family Macropodidae) typified in the high country by the Red-necked Wallaby *Macropus rufogriseus* and Black Wallaby *Wallabia bicolor*, though escapee captive Tammar Wallabies *Macropus eugenii* turn up as eagle prey.

Wallacea The Indonesian islands east of Wallace's Line, and thus part of the Australasian faunal region: Sulawesi, the Moluccas, and the Lesser Sunda Islands.

Wing-loading The amount of weight supported by a unit area of the wing, high in a duck or Peregrine, low in a Black Kite.

Zygodactyl Avian foot arrangement in which two toes (1 and 4) are facultatively (as in owls) or permanently oriented backwards and the other two (2 and 3) forward (see **Anisodactyl**).

References

ACT Land and Planning Authority (2004) *Canberra Spatial Plan*. ACT Land and Planning Authority, Canberra.

Aumann T (1988) The diet of the Brown Goshawk, *Accipiter fasciatus*, in South-eastern Australia. *Australian Wildlife Research* **15**, 587–594.

Aumann T (1989) Breeding parameters of the Brown Goshawk *Accipiter fasciatus* in south-eastern Australia. *Emu* **89**, 112–118.

Aumann T (2001) Breeding biology of raptors in riparian environments in the south-west of the Northern Territory, Australia. *Emu* **101**, 305–315.

Bader TJ, Bednarz JC (2011) Parental care and diet of Mississippi Kites (*Ictinia mississippiensis)* in eastern Arkansas. *Journal of Raptor Research* **45**, 109–118.

Baker GB, Olsen P, McCulloch R, Dettman EB (1997) Preliminary results of a long-term study of the Nankeen Kestrel: Population density and turnover. In *Australian Raptor Studies II* (Eds G Czechura and S Debus) pp. 108–113. Birds Australia Monograph 3, Birds Australia, Melbourne.

Baker-Gabb DB (1982) Comparative ecology and behaviour of Swamp Harriers *Circus approximans*, Spotted Harriers *C. assimilis* and raptors in Australia and New Zealand. PhD thesis. Monash University, Clayton.

Baker-Gabb DJ (1984) The breeding ecology of twelve species of diurnal raptor in north western Victoria. *Australian Wildlife Research* **11**, 145–160.

Baker-Gabb DJ, Fitzherbert K (1989) An overview of raptor movements and wintering places in Australia and New Zealand. In *Raptors in the Modern World* (Eds B-U Meyburg and RD Chancellor) World Working Group on Birds of Prey and Owls. Berlin, Germany.

Balgooyen TG (1976) Behavior and ecology of the American Kestrel (*Falco sparverius* L.) in the Sierra Nevada of California. *University of California Publications in Zoology* **103**, 1–85.

Balgooyen T (1989) Natural history of the American Kestrel in Venezuela. *Journal of Raptor Research* **23**, 85–93.

Barnes T (2005) Foraging, habitat use and nesting of the Black-shouldered Kite *Elanus axillaris* in the Australian Capital Territory. *Australian Field Ornithology* **22**, 58–66.

Bennett AF, Lumsden LF, Alexander JSA, Duncan PE, Johnson PG, Robertson P, Silveira CE (1991) Habitat use by aboreal marsupials along an environmental gradient in north-eastern Victoria. *Wildlife Research* **18**, 125–146.

Bilney RJ, Cooke R, White J (2006) Change in diet of Sooty Owls (*Tyto *tenebricosa*) since European settlement: From terrestrial to arboreal prey and increased overlap with Powerful Owls. *Wildlife Research* **33**, 17–24.

Bollen C (1991) Breeding behaviour and diet of the Australian Kestrel *Falco cenchroides* on the Southern Tablelands of New South Wales. *Australian Bird Watcher* **14**, 44–50.

Bollen C (1993) Breeding behaviour and diet of the Brown Falcon *Falco berigora* on the Southern Tablelands of New South Wales. *Australian Bird Watcher* **15**, 19–23.

Botehlo ES, Gennero AL, Arrowood PC (1993) Parental care, nestling behaviour and nestling interaction in a Mississippi Kite *Ictinia mississippiensis* nest. *Journal of Raptor Research* **27**, 16–20.

Bunn DS, Warburton AB, Wilson RDS (1982) *The Barn Owl.* T & AD Poyser, Calton, UK.

Burton AM, Alford RA, Young J (1994) Reproductive parameters of the Grey Goshawk (*Accipiter novaehollandiae*) and Brown Goshawk (*Accipiter fasciatus*) at Abergowrie, northern Queensland, Australia. *Journal of Zoology* **232**, 347–363.

Burton JA (Ed.) (1973) *Owls of the World: Their Evolution, Structure and Ecology.* EP Dutton, New York.

Cade TJ (1982) *Falcons of the World.* Collins, London.

Cade TJ, Enderson JH, Thelander CG, White CM (Eds) (1988) *Peregrine Falcon Populations: Their Management and Recovery.* Peregrine Fund, Boise, Idaho, USA.

Campbell B, Lack E (1985) (Eds) *A Dictionary of Birds.* T & AD Poyser, Calton, UK.

Canberra Ornithologists Group (2003) Annual bird report: 1 July 2002 to 30 June 2003. *Canberra Bird Notes* **28**, 125–188.

Canberra Ornithologists Group (2005) Annual bird report: 1 July 2003 to 30 June 2004. *Canberra Bird Notes* **30**, 1–64.

Canberra Ornithologists Group (2008) Nomination of a Vulnerable species to the ACT Flora and Fauna Committee – Little Eagle. *Canberra Bird Notes* **33**, 84–95.

Caro TM, O'Doherty G (1999) On the use of surrogate species in conservation biology. *Conservation Biology* **13**, 805–814.

Carson R (1962) *Silent Spring.* Houghton Mifflin, Boston.

Cartron JE (Ed.) (2010) *Raptors of New Mexico.* University of New Mexico Press, Albuquerque.

Cherriman SC, Foster A, Debus SJS (2009) Supplementary notes on the breeding behaviour of Wedge-tailed Eagles *Aquila audax. Australian Field Ornithology* **26**, 142–147.

Collins L, Croft DB (2007) Factors influencing chick survival in the Wedge-tailed Eagle *Aquila audax. Corella* **31**(2), 32–40.

Cooke R, Wallis R, Hogan F, White J, Webster A (2006) The diet of Powerful Owls (*Ninox strenua*) and prey availability in a continuum of habitats from disturbed urban fringe to protected forest environments in south-eastern Australia. *Wildlife Research* **33**, 199–206.

Corbett L, Hertog T (2011) Diet and breeding of White-bellied Sea-Eagles *Haliaeetus leucogaster* in the Northern Territory, Australia. *Corella* **35**, 41–48.

Craighead JJ, Craighead FC (1956) *Hawks, Owls and Wildlife*. Dover, New York.

Cramp S, Simmons KEL (Eds) (1980) *Handbook of the Birds of Europe, the Middle East and North Africa: The Birds of the Western Palearctic. Vol. II Hawks to Bustards*. Oxford University Press, Oxford.

Cupper J, Cupper L (1981) *Hawks in Focus*. Jaclin Enterprises, Mildura, Vic.

de la Torre J (1990) *Owls: Their Life and Behavior*. Crown Publishers, New York.

Debus SJS (2005) White-bellied Sea-Eagles breeding in the Australian Capital Territory? *Canberra Bird Notes* **30**, 146–147.

Debus SJS (2008) Biology and diet of the White-bellied Sea-Eagle *Haliaeetus leucogaster* breeding in northern inland New South Wales. *Australian Field Ornithology* **25**, 165–193.

Debus S (2009) *The Owls of Australia; A Field Guide to Australian Night Birds*. Envirobook, Canterbury, NSW.

Debus SJS (2011) Parental time budgets and breeding behaviour of the Little Eagle *Hieraaetus morphnoides* in northern New South Wales. *Corella* **35**, 65–72.

Debus S (2012) *Birds of Prey of Australia: A Field Guide*. CSIRO Publishing, Melbourne.

Debus SJS, Ley AJ (2009) Aspects of the breeding cycle of the Little Eagle *Hieraaetus morphnoides*. *Australian Field Ornithology* **26**, 76–99.

Debus S, Ley AJ, Trémont SM, Trémont RM (1991) Breeding behaviour and diet of the Australian Hobby *Falco longipennis* in northern New South Wales. *Australian Bird Watcher* **14**, 123–137.

Debus SJS, Ley AJ, Trémont SM, Trémont RM, Collins JL (1993) Breeding behaviour and diet of the Collared Sparrowhawk *Accipiter cirrhocephalus* in northern New South Wales. *Australian Bird Watcher* **15**, 68–91.

Debus SJS, Olsen J, Rose AB (2004) Diet of the Barn Owl *Tyto alba* near Lake Frome in arid South Australia. *Corella* **28**, 40–42.

Debus SJS, Olde GS, Marshall N, Meyer J, Rose AB (2006) Foraging, breeding behaviour and diet of a family of Black-shouldered Kites *Elanus axillaris* near Tamworth, New South Wales. *Australian Field Ornithology* **23**, 130–143.

Debus SJS, Hatfield TS, Ley AJ, Rose AB (2007a) Breeding biology and diet of the Wedge-tailed Eagle *Aquila audax* in the New England region of New South Wales. *Australian Field Ornithology* **24**, 93–120.

Debus SJS, Hatfield TS, Ley AJ, Rose AB (2007b) Breeding biology and diet of the Little Eagle *Hieraaetus morphnoides* in the New England region of New South Wales. *Australian Field Ornithology* **24**, 137–157.

Debus SJS, Olsen J, Judge D, Butterfield M (2013) Numbers of breeding Little Eagles *Hieraaetus morphnoides* near Canberra in relation to atlas counts. *Corella* **37**, 30–32.

Dennis T, McIntosh R, Shaughnessy P (2011) Effects of human disturbance on productivity of White-bellied Sea-Eagles, *Haliaeetus leucogaster*. *Emu* **111**, 179–185.

Dijkstra C, Bult A, Bijlsma S, Dann S, Meijer T, Zilstra M (1990) Brood size manipulations in the Kestrel *Falco tinnunculus*: effects on offspring and parent survival. *Journal of Animal Ecology* **59**, 269–286.

Duncan J (2003) *Owls of the World: Their Lives, Behavior and Survival.* Firefly Books, Buffalo, NY.

Dunk JR (1995) White-tailed Kite (*Elanus leucurus*), *The Birds of North America Online* (Ed. A Poole). Ithaca, Cornell Lab of Ornithology, Retrieved from the Birds of North America Online: <http://bna.birds.cornell.edu/bna/species/178>.

Edwards GP, Dobbie W, Berman DMcK (2002a) Warren ripping: its impacts on wild European rabbits and other wildlife amid the establishment of rabbit haemorrhagic disease. *Wildlife Research* **29**, 567–575.

Edwards GP, Dobbie W, Berman DMcK (2002b) Population trends in European rabbits and other wildlife of central Australia in the wake of rabbit haemorrhagic disease. *Wildlife Research* **29**, 557–565.

Ellis DH, Nelson RW (2010) Cheaper by the dozen: Bald Eagles capture many fingerlings with each attempt at fish boils in coastal British Colombia. *Journal of Raptor Research* **44**, 151–154.

Emison WB, White CM, Hurley VG, Brimm DJ (1997) Factors influencing the breeding distribution of the Peregrine Falcon in Victoria, Australia. *Wildlife Research* **24**, 433–444.

Emison WB, Hurley VG, White CM, Brimm DJ (1998) Results from a banding study of Peregrine Falcon chicks in Victoria, 1972–1997. *Corella* **22**, 87–91.

Encyclopaedia Britannica World Atlas (2006) Encyclopaedia Britannica, Inc., London.

Ferguson-Lees J, Christie DA (2001) *Raptors of the World.* Christopher Helm, London.

Fleay D (1968) *Nightwatchmen of Bush and Plain.* Jacaranda, Milton, Queensland.

Foster A, Wallis R (2010) Nest-site characteristics of the Wedge-tailed Eagle *Aquila audax* in southern Victoria. *Corella* **34**, 36–44.

Fox NC (1977) The biology of the New Zealand Falcon *Falco novaeseelandiae.* PhD dissertation, University of Canterbury.

Frith HJ (Ed.) (1969) *Birds in the Australian High Country.* Reed, Sydney.

Fuentes E, Olsen J, Rose AB (2005) Breeding diet at two Whistling Kite nests near Canberra. *Australian Field Ornithology* **22**, 122–125.

Fuentes E, Olsen J, Rose AB (2007) Diet, occupancy, and breeding performance of Wedge-tailed Eagles *Aquila audax* near Canberra, Australia 2002–2003: four decades after Leopold and Wolfe. *Corella* **31**, 65–72.

Gammage B (2012) *The Biggest Estate on Earth: How Aborigines Made Australia.* Allen & Unwin, Crows Nest, NSW.

Gard NW, Bird DM (1990) Breeding behaviour of American Kestrels raising manipulated broods in years of varying prey abundance. *Wilson Bulletin* **102**, 605–614.

Garnett ST, Szabo JK, Dutson G (2011) *The Action Plan for Australian Birds 2010.* CSIRO Publishing, Melbourne.

Gessaman JA (1980) An evaluation of heart rate as an indirect measure of daily energy metabolism of the American Kestrel. *Comparative Biochemistry and Physiology* **65A**, 273–289.

Gittleman JL, Funk SM, Macdonald D, Wayne RK (2001) *Carnivore Conservation.* Cambridge University Press, Cambridge, UK.

Gjershaug JO, Lerner HRL, Diserud OH (2009) Taxonomy and distribution of the Pygmy Eagle *Aquila (Hieraaetus) weiskei* (Accipitriformes: Accipitridae). *Zootaxa* **2326**, 24–38.

Glinski RL, Ohmart RD (1983) Breeding ecology of the Mississippi Kite in Arizona. *Condor* **85**, 200–207.

Goslow CE (1971) The attack and strike of some North American raptors. *The Auk* **88**, 815–827.

Grossman ML, Hamlet J (1964) *Birds of Prey of the World.* Bonanza Books, Clarkson, New York.

Hagen Y (1969) Norwegian studies on the reproduction of birds of prey and owls in relation to micro-rodent population fluctuations. *Fauna* **22**, 73–126.

Heinrich B (1987) *One Man's Owl.* Princeton University Press, Princeton, NJ.

Henzell RP, Cunningham RB, Neave HM (2002) Factors affecting the survival of Australian wild rabbits exposed to rabbit haemorrhagic disease. *Wildlife Research* **29**, 523–542.

Higgins PJ (Ed.) (1999) *Handbook of Australian, New Zealand and Antarctic Birds*, Vol. 4. Oxford University Press, Melbourne.

Hollands D (1984) *Eagles, Hawks and Falcons of Australia.* Nelson, Melbourne.

Hollands D (2008) *Owls, Frogmouths and Nightjars of Australia.* Bloomings Books, Melbourne.

James JW (1980) Food of the Powerful Owl *Ninox strenua* in south-eastern Queensland. *Emu* **80**, 34–35.

Johnsgard PA (1990) *Hawks, Eagles & Falcons of North America.* Smithsonian Institution, Washington DC, USA.

Johnsgard PA (2002) *North American Owls: Biology and Natural History.* 2nd edn. Smithsonian Institution Press, Washington DC.

Kavanagh R (2002a) Comparative diets of the Powerful Owl *Ninox strenua*, Sooty Owl *Tyto tenebricosa* and Masked Owl *Tyto novaehollandiae* in southeastern Australia. In *The Ecology and Conservation of Owls.* (Eds I Newton, R Kavanagh, J Olsen and I Taylor) pp 175–191. CSIRO Publishing, Melbourne.

Kavanagh R (2002b) Conservation and management of large forest owls in southeast Australia. In *The Ecology and Conservation of Owls.* (Eds I Newton, R Kavanagh, J Olsen and I Taylor) pp. 201–219. CSIRO Publishing, Melbourne.

Kavanagh RP, Bamkin KL (1995) Distribution of nocturnal birds and mammals in relation to the logging mosaic in south-eastern New South

Wales, Australia. *Biological Conservation* **71**, 41–53.
Kenward RE 1990 Feeding habits. In *Birds of Prey* (Ed. I Newton) pp. 90–107. Merehurst, London.
Kirmse W (1994) Raptors' plasticity of nest site selection. In *Raptor Conservation Today.* (Eds B-U Meyburg and RD Chancellor) pp. 143–145. World Working Group of Birds of Prey and Owls, The Pica Press, London.
Kirmse W (2004) Tree nesting Peregrines *Falco p. peregrinus* in Europe did not recover. In *Raptors Worldwide.* (Eds RD Chancellor and B-U Meyburg) pp. 271–277. World Working Group on Birds of Prey and Owls Berlin, and Birdlife Hungary, Budapest.
König C, Weick F (2008) *Owls of the World.* 2nd edn. Christopher Helm, London.
Konishi M (1973) How the owl tracks its prey. *American Scientist* **61**, 414–424.
Konishi M (1983) Night owls are good listeners. *Natural History* **92**(9), 56–59.
Konishi M (1993) Listening with two ears. *Scientific American* **268**(4), 66–73.
Körner C, Paulsen J (2004) A worldwide study of high altitude treeline temperatures. *Journal of Biogeography* **31**, 713–732.
Krackow S (1993) Note on the falconiforme sex ratios given in Olsen and Cockburn 1991: Avian raptors exhibit no unique sex-bias. *Behavioral Ecology and Sociobiology* **32**, 429–430.
Krebs C (2008) *The Ecological World View.* CSIRO Publishing, Melbourne.
Krüger O (2005) The evolution of reversed sexual dimorphism in hawks, falcons and owls: A comparative study. *Evolutionary Ecology* **19**, 467–486.
Lansell GL (1936) A few more notes on Wedge-tailed Eagles and notes on ravens. *Emu* **36**, 246–247.
Lasiewski RC, Dawson WR (1967) A re-examination of the relation between standard metabolic rate and body weight in birds. *Condor* **69**, 13–23.
Lavazanian E, Wallis R, Webster A (1994) Diet of Powerful Owls (*Ninox strenua*) living near Melbourne, Victoria. *Wildlife Research* **21**, 643–646.
Leopold AS, Wolfe TO (1970) Food habits of the Wedge-tailed Eagles, *Aquila audax*, in South-eastern Australia. *CSIRO Wildlife Research* **15**, 1–17.
Liébana MS, Sarasola JH, Bo MS (2009) Parental care and behavior of breeding American Kestrels (*Falco sparverius*) in central Argentina. *Journal of Raptor Research* **43**, 338–344.
Lindenmayer DB, Cunningham RB, Pope M, Donnelly CF (1999) The Tumut fragmentation experiment in south-eastern Australia: The effects of landscape context and fragmentation on arboreal marsupials. *Ecological Applications* **9**, 594–611.
Mallinson D, Olsen P, Olsen J (1990) A note on the breeding biology of the Little Eagle *Hieraaetus morphnoides* and other raptors in the Mount Mugga Area, ACT. *Corella* **14**, 88–91.
Marchant S, Higgins PJ (Eds) (1993) *Handbook of Australian, New Zealand and Antarctic Birds* Vol. 2, Oxford University Press, Melbourne.
Marks JS, Cannings RJ, Mikkola H (1999) Family Strigidae (typical owls).

In *Handbook of Birds of the World. Vol. 5. Barn Owls to Hummingbirds.* (Eds J del Hoyo, A Elliott and L Sargatal) pp. 76–151. Lynx Edicions, Barcelona.

Marti CD (1973) Food consumption and pellet formation rates in four owl species. *Wilson Bulletin* **85**, 178–181.

Marti CD, Korpimäki E, Jaksić FM (1993) Trophic structure of raptor communities: A three-continent comparison and synthesis. In *Current Ornithology.* Vol. 10. (Ed. DM Power) pp. 47–137. Plenum Press, New York.

Martin GR, Kirkpatrick WE, King DK, Robertson ID, Hood PJ, Sutherland JR (1994) Assessment of the potential toxicity of an anticoagulant, Pindone (2-pivalyl, 3-indandione), to some Australian birds. *Wildlife Research* **21**, 85–93.

Mason R (undated) *Vegetation in the Australian Alps.* Australian Alpine Parks Educational Resource: Vegetation. Department of Sustainability, Environment, Water, Population and Communities, Canberra. <www.australianalps.environment.gov.au/learn/>.

Massemin S, Korpimaki E, Wiehn J (2000) Reversed sexual size dimorphism in raptors: Evaluation of the hypotheses in kestrels breeding in a temporally changing environment. *Oecologia* **124**, 26–32.

McDonald P (2003a) Variable plumage and bare part colouration in the Brown Falcon *Falco berigora*: the influence of age and sex. *Emu* **103**, 21–28.

McDonald PG (2003b) Nestling growth and development in the Brown Falcon, *Falco berigora*: An improved ageing formula and field-based method of sex determination. *Wildlife Research* **30**, 411–418.

McDonald PG, Olsen PD, Baker-Gabb DJ (2003) Territory fidelity, reproductive success and prey choice in the brown falcon, *Falco berigora*: A flexible bet-hedger? *Australian Journal of Zoology* **51**, 399–414.

McDonald PG, Olsen J, Rose AB (2012) The diet of breeding Brown Falcons (*Falco berigora*) in the Canberra region. *Journal of Raptor Research* **46**(4), 394–400.

McGilp JN (1925) Wedge-tailed Eagles destroying lambs. *Emu* **25**, 120.

McNabb EG (1994) The successful rehabilitation of two Powerful Owl fledglings. *Australian Bird Watcher* **15**, 287–297.

McNabb EG (1996) Observations on the biology of the Powerful Owl *Ninox strenua* in southern Victoria. *Australian Bird Watcher* **16**, 267–295.

Mendelsohn JM (1986) Sexual size dimorphism and roles in raptors – fat females, agile males. *Durban Museum Novitates* **13**, 323–336.

Menkhorst P, Knight F (2001) *A Guide to the Mammals of Australia.* Oxford University Press, Melbourne.

Metcalf EC (1989) The breeding biology of the Australian Hobby *Falco longipennis. Australian Bird Watcher* **13**, 20–29.

Mikkola H (1983) *Owls of Europe.* T & AD Poyser, Calton, UK.

Mooney N, Brothers N (1988) Tasmania. Cited in Olsen PD & Olsen, J Population Trends, Distribution and Status of the Peregrine Falcon in Australia, pp. 255–274. In *Peregrine Falcon Populations: Their Management and*

Recovery. (Eds TJ Cade, JH Enderson, CG Thelander and CM White). Peregrine Fund, Boise, Idaho.

Mosher JA, Matray PF (1974) Size dimorphism: A factor in energy savings for Broad-winged Hawks. *The Auk* **91**, 325–341.

Murdoch WW, Oaten A (1975) Predation and population stability. *Advances in Ecological Research* **9**, 1–131.

Negro JJ, Pertoldi C, Randi E, Ferrero JJ, Lopez-Caballero JM, Rivera D, Korpimaki E (2006) Convergent evolution of the *Elanus* kites and the owls. *Journal of Raptor Research* **40**, 222–225.

Nelson RW (1977) Behavioral Ecology of Coastal Peregrines *Falco peregrinus pealei*. PhD thesis University of Calgary, Alberta.

Newsome A, Pech R, Smyth R, Banks P, Dickman C (1997) 'Potential impacts on Australian native fauna of rabbit calicivirus disease'. Report to the Australian Nature Conservation Agency, Canberra.

Newton I (1979) *Population Ecology of Raptors.* T & AD Poyser, Berkhamsted UK.

Newton I (1986) *The Sparrowhawk.* T & AD Poyser, Calton, UK.

Newton I (1998) *Population Limitation in Birds.* Academic Press, London.

Norris KC, Emison WB, Bren WB (1977) A preliminary survey of the population of Peregrine Falcons in Victoria. *Emu* **77**, 86–87.

Olsen J (1989) Some thoughts on risk of accidents in raptors. *Australasian Raptor Association News* **10**, 69–72.

Olsen J (1990) *Caring for Birds of Prey.* Applied Ecology Research Group, Canberra.

Olsen J (1992) 'Raptors in Namadgi, Canberra Nature Parks, the Murrumbidgee River Corridor, and Googong Foreshore'. Report to ACT Parks and Conservation Service.

Olsen J (1995) *Some Time with Eagles and Falcons.* Hancock House, Surrey, BC, Canada.

Olsen J (2011a) *Australian High Country Owls.* CSIRO Publishing, Melbourne.

Olsen J (2011b) Does upside-down abundance of possums and gliders make male Powerful Owls big? *Boobook* **29**, 49–50.

Olsen J (2013) Reversed sexual dimorphism and prey size taken by male and female raptors: a comment on Pande and Dahanukar (2012). *Journal of Raptor Research* **47**, 79–81.

Olsen J, Georges A (1993) Do Peregrine fledglings reach independence during peak abundance of their main prey? *Journal of Raptor Research* **27**, 149–153.

Olsen J, Olsen P (1980a) Observations on defence of the nest against humans by Australian *Falco. Emu* **80**, 163–165.

Olsen J, Rehwinkel R (1995) 'Peregrines and Powerful Owls in Namadgi and Tidbinbilla A.C.T.' [National Estate Grant 1993–1995 103 page report].

Olsen J, Trost S (2007) Duelling and nest failure in Southern Boobooks *Ninox novaeseelandiae. Australian Field Ornithology* **24**, 13–25.

Olsen J, Trost S, Hayes G (2002) Vocalisations used by Southern Boobooks *Ninox novaeseelandiae* in and adjoining Aranda Bushland, Canberra, Australia. In *The Ecology and Conservation of Owls.* (Eds I Newton, R Kavanagh, J Olsen and I Taylor) pp. 305–319. CSIRO Publishing, Melbourne.

Olsen J, Tucker AD (2003) A brood-size manipulation experiment with Peregrine Falcons, *Falco peregrinus*, near Canberra. *Emu* **103**, 127–132.

Olsen J, Debus SJS, Rose AB, Hayes G (2004) Breeding success, cliff characteristics and diet of Peregrine Falcons at high altitude in the Australian Capital Territory. *Corella* **28**, 33–37.

Olsen J, Fuentes E, Rose AB (2006a) Male Peregrine Falcon *Falco peregrinus macropus* fledged from a cliff nest found breeding in a stick nest. *Australian Field Ornithology* **23**, 8–14.

Olsen J, Fuentes E, Rose AB, Trost S (2006b) Food and hunting of eight breeding raptors near Canberra, 1990–1994. *Australian Field Ornithology* **23**, 77–95.

Olsen J, Fuentes E, Rose AB (2006c) Trophic relationships between neighbouring White-bellied Sea-Eagles *Haliaeetus leucogaster* and Wedge-tailed Eagles *Aquila audax* breeding on rivers and dams near Canberra. *Emu* **106**, 193–201.

Olsen J, Fuentes E, Bird DM, Rose AB, Judge D (2008) Dietary shifts based on prey availability in Peregrine Falcons and Australian Hobbies near Canberra, Australia. *Journal of Raptor Research* **42**, 125–137.

Olsen J, Judge D, Fuentes E, Rose AB, Debus S (2010) Diets of Wedge-tailed Eagles (*Aquila audax*) and Little Eagles (*Hieraaetus morphnoides*) breeding near Canberra, Australia. *Journal of Raptor Research* **44**, 50–61.

Olsen J, Osgood M, Maconachie M, Dabb G (2012) Little Eagles, Whistling Kites and Swamp Harriers in the Australian Capital Territory 2011. *Canberra Bird Notes* **37**(3), 206–209.

Olsen J, Debus SJS, Judge D (2013a) Declining Little Eagles *Hieraaetus morphnoides* and increasing rabbit numbers near Canberra: Is secondary poisoning by Pindone the problem? *Corella* **37**, 33–35.

Olsen J, Debus SJS, Judge D, Rose AB (2013b) Diets of Wedge-tailed Eagles *Aquila audax* and Little Eagles *Hieraaetus morphnoides* breeding near Canberra, 2008–2009. *Corella* **37**, 25–29.

Olsen J, Judge D, Trost S, Rose AB (2013c) Do inverse population densities of Possums and Gliders determine sexual dimorphism in Powerful Owls? *Australian Field Ornithology* **30**, 22–39.

Olsen J, Debus SJS, Judge D, Rose AB (2013d) Diets of White-bellied Sea-Eagles *Haliaeetus leucogaster* and Whistling Kites *Haliastur sphenurus* breeding near Canberra, 2003–2008. *Corella* **37**, 13–18.

Olsen P, Olsen J (1979) Eggshell thinning in the peregrine *Falco peregrinus* (Aves: Falconidae) in Australia. *Australian Wildlife Research* **6**, 217–226.

Olsen P, Olsen J (1980b) Observations on development, nestling chronology, and clutch and brood size in the Australian Kestrel *Falco cenchroides* (Aves: Falconidae). *Emu* **7**, 247–255.

Olsen P, Debus SJS, Czechura GV, Mooney NJ (1990) Comparative feeding ecology of the Grey Goshawk *Accipiter novaehollandiae* and Brown Goshawk *Accipiter fasciatus*. *Australian Bird Watcher* **13**, 178–192.

Olsen P, Olsen J, Mason I (1993) Breeding and non-breeding diet of the Peregrine Falcon *Falco peregrinus* near Canberra, prey selection, and the relationship between diet and reproductive success. In *Australian Raptor Studies*. (Ed. PD Olsen) pp. 55–77. RAOU, Melbourne.

Olsen PD, Marples TG (1992) Alteration in clutch size in raptors in response to changes in prey availability: evidence from control of a broad-scale rabbit infestation. *Wildlife Research* **19**, 129–135.

Olsen PD, Olsen J (1987) Estimating the age of nestling raptors. *Australian Bird Watcher* **12**, 130–131.

Olsen PD, Olsen J (1988) Population trends, distribution and status of the Peregrine Falcon in Australia. In *Peregrine Falcon Populations: Their Management and Recovery*. (Eds TJ Cade, JH Enderson, CG Thelander and CM White) pp. 255–274. The Peregrine Fund, Boise, Idaho.

Parker BD, Hume ID, Boles WE (2007) Diet of breeding Wedge-tailed Eagles *Aquila audax* in south-central Queensland. *Corella* **31**, 50–62.

Paull D (1991) Foraging and breeding behaviour of the Australian Kestrel *Falco cenchroides* on the Northern tablelands of New South Wales. *Australian Bird Watcher* **14**, 85–92.

Peeters H (2007) *Owls of California and the West*. University of California Press, Berkeley and Los Angeles.

Peeters H, Peeters P (2005) *Raptors of California*. University of California Press, Berkeley and Los Angeles.

Perrins C (Ed.) (2009) *The Princeton Encyclopedia of Birds*. Princeton University Press, Princeton, New Jersey.

Poole AF (1989) *Ospreys: A Natural and Unnatural History*. Cambridge University Press, Cambridge, UK.

Ratcliffe D (1993) *The Peregrine Falcon*. 2nd edn. Poyser, Calton, UK.

Ratcliffe D (1997) *The Raven*. Poyser, London.

Ray JC (2005) Large carnivorous animals as tools for conserving biodiversity: Assumptions and uncertainties. In *Large Carnivores and the Conservation of Biodiversity*. (Eds JC Ray, K Redford, R Steneck and J Berger) pp. 34–56. Island Press, Washington DC.

Ray JC, Redford KH, Steneck RS, Berger J (Eds) (2005) *Large Carnivores and the Conservation of Biodiversity*. Island Press, Washington DC.

Read DG (2005) Observations on parent/young behaviour in a pair of Black-shouldered Kites *Elanus axillaris*. *Australian Field Ornithology* **22**, 109–121.

Real J (1996) Biases in diet study methods in the Bonelli's Eagle. *Journal of Wildlife Management* **60**, 632–638.

Reardon M (2003) Too many roos: Our national emblem may be eating us out of house and home. *Australian Geographer* **73**, 86–97.

Reynolds RT (1972) Sexual dimorphism in accipiter hawks: A new hypothesis. *Condor* **74**, 191–197.

Riddell W (2011a) The juvenile plumage of the Grey Goshawk *Accipiter novaehollandiae* in tropical Australia. *Australian Field Ornithology* **28**, 180–185.

Riddell W (2011b) Double-brooding by the Brown Goshawk *Accipiter fascia-*

tus in the Northern Territory. *Australian Field Ornithology* **28**, 92–93.

Ridpath MG, Brooker MG (1986) The breeding of the Wedge-tailed Eagle *Aquila audax* in relation to its food supply in arid Western Australia. *Ibis* **128**, 177–194.

Robertson G (1987) Effects of drought on a breeding population of Wedge-tailed Eagles *Aquila audax*. *Emu* **87**, 220–223.

Rohner C, Smith JN (1996) Brood size manipulations in Great Horned Owls *Bubo virginianus*: are predators food limited at the peak of prey cycles? *Ibis* **138**, 236–242.

Sandell PR, Start AN (1999) 'Rabbit Calicivirus Disease Program Report 4: Implications for biodiversity in Australia'. A report of research conducted by participants of the Rabbit Calicivirus Disease Monitoring Program and Epidemiology Research Program. Prepared for the RCD Management Group. Bureau of Rural Sciences, Canberra.

Saunders G, Choquenot D, McIlroy J, Packwood R (1999) Initial effects of rabbit haemorrhagic disease on free-living rabbit (*Oryctolagus cuniculus*) populations in central-western New South Wales. *Wildlife Research* **26**, 69–74.

Schodde D, Mason I (1980) *Nocturnal Birds of Australia*. Lansdowne, Melbourne.

Seebeck JH (1976) The diet of the Powerful Owl *Ninox strenua* in western Victoria. *Emu* **76**, 167–170.

Seguin JF, Bayle P, Thibault JC, Torre J, Vigne JD (1998) A comparison of methods to evaluate the diet of the Golden Eagle in Corsica. *Journal of Raptor Research* **32**, 314–318.

Sharp A, Gibson L, Norton M, Marks A, Ryan B, Semeraro L (2002a) An evaluation of the use of regurgitated pellets and skeletal material to quantify the diet of Wedge-tailed Eagles, *Aquila audax*. *Emu* **102**, 181–185.

Sharp A, Gibson L, Norton M, Ryan B, Marks A, Semeraro L (2002b) The breeding season diet of the Wedge-tailed Eagle (*Aquila audax*) in western New South Wales and the influence of Rabbit Calicivirus disease. *Wildlife Research* **29**, 175–184.

Shephard JM, Catterall CP, Hughes JM (2005) Long-term variation in the distribution of the White-bellied Sea-Eagle *Haliaeetus leucogaster* across Australia. *Austral Ecology* **30**, 131–145.

Sherrod S (1983) *Behaviour of Fledged Peregrines*. Peregrine Fund, Ithaca, New York.

Silva LM, Croft DB (2007) Nest-site selection, diet and parental care of the Wedge-tailed Eagle *Aquila audax* in western New South Wales. *Corella* **31**, 23–31.

Simmons R (1988) Honest advertising, sexual selection, courtship displays, and body condition of polygynous male harriers. *The Auk* **105**, 303–307.

Southey T (1848) *The Rise, Progress and Present State of Colonial Wools*. Smith, Elder and Co., London.

Stanton M (2011) Barking Owl diet in the Pillaga Forests of northern New South Wales. MSc Thesis. University of New England, Armidale, NSW.

Starr MJ, Starr M, Wilson SC (2004) Hunting rates and prey of a pair of breeding Nankeen Kestrels *Falco cenchroides* near Sydney, New South Wales. *Australian Field Ornithology* **21**, 71–75.

Steele W, Baker-Gabb D (2009) A national community-based survey of the diurnal birds of prey (BOP Watch). *Boobook* **27**, 23–24. [Abstract.]

Steenhof K, Kochert MN (1988) Dietary responses of three raptor species to changing prey densities in a natural environment. *Journal of Animal Ecology* **57**, 37–48.

Storer RW (1966) Sexual dimorphism and food habits in three North American accipiters. *The Auk* **83**, 423–436.

Strive T, Wright J, Kovaliski J, Botti G, Capucci L (2010) The non-pathogenic Australian lagovirus RCV-A1 causes a prolonged infection and elicits partial cross-protection to rabbit haemorrhagic disease virus. *Virology* **398**, 125–134.

Sunde P, Bølstad MS, Møller JD (2003) Reversed sexual dimorphism in tawny owls, *Strix aluco*, correlates with duty division in breeding effort. *Oikos* **101**, 265–278.

Tarboton WR (1978) Hunting and energy budget of the Black-shouldered Kite. *Condor* **80**, 88–91.

Thomas RH, Choudhari MM, Joslin RD (2002) *Flow and Noise Control: Review and Assessment of Future Directions.* NASA-TM-2002-211631. National Aeronautics and Space Administration, Langley Research Station, Hampton, VA, USA.

Tilley S (1982) The diet of the Powerful Owl *Ninox strenua* in Victoria. *Australian Wildlife Research* **9**, 157–175.

Trost S, Olsen J, Rose AB, Debus SJS (2008) Winter diet of Southern Boobooks *Ninox novaeseelandiae* in Canberra 1997–2005. *Corella* **32**, 66–70.

Tucker VA (1992) Pitching equilibrium, wing span and tail span in a gliding Harris' Hawk, *Parabuteo unicinctus. Journal of Experimental Biology* **165**, 21–41.

Tucker VA (1993) Gliding birds: Reduction of induced drag by wing tip slots between the primary feathers. *Journal of Experimental Biology* **180**, 285–310.

Turner GW (Ed.) (1987) *The Australian Concise Oxford Dictionary.* Oxford University Press, Melbourne.

Van Dyck S, Strahan R (Eds) (2008) *The Mammals of Australia,* 3rd edn. Reed New Holland, Sydney.

Veerman PA (2002) *Canberra Birds: A Report on the First 21 Years of the Garden Bird Survey.* Author, Canberra.

Walker RB (1963) The economic development of New England in the nineteenth century. In *New England Essays*. pp. 75–84. University of New England, Armidale, NSW.

Walter H (1979) *Eleonora's Falcon: Adaptations to Prey and Habitat in a Social Raptor.* University of Chicago Press, Chicago.

Ward SJ (2000) The efficacy of nestboxes versus spotlighting for detecting Feathertail Gliders. *Wildlife Research* **27**, 75–79.

Weatherly R, Baker-Gabb DJ, Mooney NJ (1985) Juvenile plumage variation in the Brown Falcon *Falco berigora*. *Emu* **85**, 57–260.

Welty JC (1982) *The Life of Birds*. 3rd edn. WB Saunders, Philadelphia.

Westgarth W (1853) *Victoria: Late Australia Felix, Port Phillip District of New South Wales*. Oliver and Boyd, London.

Winkel P (2007) Feeding ecology of the Wedge-tailed Eagle *Aquila audax* in northwest Queensland: Interactions with lambs. *Corella* **31**, 41–49.

Wintle BA, Kavanagh RP, McCarthy MA, Burgmann MA (2005) Estimating and dealing with the detectability in occupancy surveys for forest owls and arboreal marsupials. *Journal of Wildlife Management* **69**, 905–917.

Woodford MH (1960) *A Manual of Falconry*. Black, London.

Index

www.ingramcontent.com/pod-product-compliance
Lightning Source LLC
LaVergne TN
LVHW070848210926
848482LV00016B/894
9780643109162